Geometric and Functional Inequalities and Recent Topics in Nonlinear PDEs

CONTEMPORARY MATHEMATICS

781

Geometric and Functional Inequalities and Recent Topics in Nonlinear PDEs

Virtual Conference
Geometric and Functional Inequalities and Recent Topics in Nonlinear PDEs
February 28–March 1, 2021
Purdue University, West Lafayette, IN

Emanuel Indrei
Diego Marcon
Levon Nurbekyan
Editors

AMS American Mathematical Society

2020 *Mathematics Subject Classification.* Primary 35-XX, 49-XX, 34-XX, 31-XX, 53-XX, 45-XX, 42-XX, 51-XX, 52-XX.

Library of Congress Cataloging-in-Publication Data

Indrei, Emanuel, 1984– editor. | Marcon, Diego, 1986– editor. | Nurbekyan, Levon, 1986– editor. | Geometric and Functional Inequalities and Recent Topics In Nonlinear PDEs (Conference) (2021 : Purdue University, West Lafayette, Ind.)

Title: Geometric and functional inequalities and recent topics in nonlinear PDEs / [edited by] Emanuel Indrei, Diego Marcon, Levon Nurbekyan.

Description: Providence, Rhode Island : American Mathematical Society, [2023] | Series: Contemporary mathematics, ISSN: 0271-4132 (print); ISSN: 1098-3627 (online); 781 | "Virtual Conference on Geometric and Functional Inequalities and Recent Topics in Nonlinear PDEs, February 28 – March 1, 2021, Purdue University, West Lafayette, IN." | Includes bibliographical references.

Identifiers: LCCN 2022032626 | ISBN 9781470466527 (paperback) | ISBN 9781470472085 (ebook)

Subjects: LCSH: Differential equations, Partial–Congresses. | Differential equations, Nonlinear–Congresses. | Differential inequalities–Congresses. | Geometry, Differential–Congresses. | AMS: Partial differential equations. | Calculus of variations and optimal control; optimization. | Ordinary differential equations. | Potential theory. | Differential geometry. | Integral equations. | Harmonic analysis on Euclidean spaces. | Geometry. | Convex and discrete geometry.

Classification: LCC QA374 .G383 2023 | DDC 515/.353–dc23/eng20221206

LC record available at https://lccn.loc.gov/2022032626

DOI: https://doi.org/10.1090/conm/781

∞ The paper used in this book is acid-free and falls within the guidelines
established to ensure permanence and durability.
Visit the AMS home page at `https://www.ams.org/`

10 9 8 7 6 5 4 3 2 1   23 22 21 20 19 18

Contents

Preface

Historically, the most difficult problems are at the intersection of several fields of mathematics. The mathematical content of this volume is at the intersection of viscosity theory, Fourier analysis, mass transport theory, fractional elliptic theory, and geometric analysis. Recently, advanced tools have emerged which successfully and elegantly deal with long-standing problems. The most ubiquitous example in the last two decades is optimal transport theory which addressed sharp estimates in a concise manner through the regularity theory for the Monge-Ampère equation via the quadratic cost. Interestingly, the regularity theory for the classical 1781 Monge cost is still not fully developed. However, in several applications, other cost functions appear and a regularity theory to address this was developed via the Ma-Trudinger-Wang (MTW) condition at the start of the 21st century.

The 19th century witnessed the development of the theory of electromagnetism via Maxwell's equations inclusive of Ampère's law, Faraday's law, and Gauss' law. One of the features is the propagation of electromagnetic waves: light from the sun, radio waves, computer signals, etc. The central partial differential equation (PDE) in the theory is the wave equation and there are several still unresolved estimates today. The classical tool to address this is the Fourier transform.

In a first complex analysis class, Liouville's theorem appears: an entire and bounded function is constant. In modern PDE theory several Liouville-type theorems are proved for fully nonlinear elliptic equations and there are many open problems which naturally emerge. An interesting regularity theory which permits weak solutions in an elegant geometric way is encoded in viscosity theory. An application of the theory also includes isoperimetric-type inequalities.

A historically ubiquitous inequality is the classical isoperimetric inequality which is summed up via: under a fixed mass constraint the Euclidean ball minimizes the isotropic perimeter in the class of sets of finite perimeter. Ironically, if one includes a potential energy in the total energy, the isoperimetric problem is not fully understood although several advances appeared in recent years. One characteristic which emerges is convexity and the way a minimizer changes via the change in mass is of central interest. Supposing one considers the anisotropic perimeter without a potential energy, the Wulff shape emerges as the minimizer. The applications involve crystal formation and recently fractional elliptic regularity theory considered crystal defects via shearing forces and edge dislocations. A similar problem involves fixing the width and identifying shapes which maximize/minimize the volume. Historically, this was considered by individuals such as Lebesgue and Blaschke. Novel tools are therefore necessary to address these problems and of interest are geometric flows which generate a convergence into a well-known object, e.g. a ball.

A glimpse into these fields of mathematics appears in this volume:

- *A deterministic counterexample for high dimensional L^2L^∞ Strichartz estimates for the Wave equation* by Cristian Gavrus answers the question of homogeneous L^2L^∞ Strichartz estimates for the wave equation in dimensions $n \geq 4$ raised by Fang and Wang via a deterministic example.
- *On the Liouville property for fully nonlinear equations with superlinear first-order terms* by Marco Cirant and Alessandro Goffi addresses the classical Liouville property for various classes of fully nonlinear PDEs, emphasizing uniformly elliptic inequalities, in the framework of viscosity solutions, with superlinear growth in the gradient variable of the type $F(x, D^2u) \geq H_i(x, u, Du)$ in $\mathbb{R}^N$, for model cases $H_1(u, Du) = u^q + |Du|^\gamma$, $H_2(u, Du) = u^q|Du|^\gamma$, and $H_3(x, u, Du) = \pm u^q|Du|^\gamma - b(x) \cdot Du$, where $q \geq 0$, $\gamma > 1$, and b is a suitable velocity field. The paper contains an in-depth discussion of diverse techniques for studying Liouville properties and includes new sharp results, counterexamples, conjectures, and numerous challenging open problems.
- *Regularity of the solution to the principal-agent problem* by Shibing Chen proves the C^1 regularity of the solution to the principal-agent problem with general preference function satisfying structural conditions which involve the MTW condition in optimal transport theory via adapting a method of Carlier and Lachand-Robert.
- *A doubly monotone flow for constant width bodies in $\mathbb{R}^3$* by Ryan Hynd discusses a classical conjecture: a constant width body is a compact, convex subset of Euclidean space in which parallel supporting planes are separated by the same distance in every direction (this distance is called its width). In $\mathbb{R}^2$, it was proved independently by Lebesgue and Blaschke over a century ago that Reuleaux triangles enclose the least amount of area (a Reuleaux triangle is the intersection of three closed disks of radius 1 which are centered at the vertices of an equilateral triangle of side length one). There have been many subsequent proofs of the Lebesgue-Blasckhe theorem. Harrell showed that Reuleaux triangles are uniquely area-minimizing among constant width shapes. In dimensions larger than two, it is known that volume-minimizing constant width bodies exist via Blaschke's selection theorem. Surprisingly little is known about these shapes. However, there are conjectured volume-minimizing constant width bodies that Meissner and Schilling constructed which are based on a regular tetrahedron, analogous to how the Reuleaux triangle is based on an equilateral triangle. Hynd introduces a flow in the space of constant width bodies in $\mathbb{R}^3$ that simultaneously increases the volume and decreases the circumradius of the shape as time increases. Starting from any initial constant width figure, he shows that the flow exists for all positive times and converges to a closed ball as time tends to infinity. Therefore this generates a technique to study the shape as the volume decreases via a time-reversal.
- *A fractional glance to the theory of edge dislocations* by Serena Dipierro, Stefania Patrizi, and Enrico Valdinoci studies the edge dislocations problem for crystals guided by the classical Peierls-Nabarro model that describes the equilibrium configurations by a nonlocal fractional Laplacian equation $-\sqrt{-\Delta}u = W'(u)$ in $\mathbb{R}$, where u is the dislocation function

along the slip line, and W is a multiwell potential penalizing the dislocated atoms. The detailed analysis of the model reveals the emergence of chaotic stationary behaviors in the form of complex and unusual equilibrium patterns. Further results address the dislocation dynamics, collisions between dislocation points, Orowan's Law, and the homogenization problem related to the Peierls-Nabarro model.

Our hope is that you will enjoy the volume!

Contemporary Mathematics
Volume **781**, 2023
https://doi.org/10.1090/conm/781/15706

A deterministic counterexample for high dimensional L^2L^∞ Strichartz estimates for the Wave equation

Cristian Gavrus

Abstract. In this note we discuss the question of homogeneous L^2L^∞ Strichartz estimates for the Wave equation in dimensions $n \geq 4$ raised by Fang and Wang and recently shown to fail by Guo, Li, Nakanishi and Yan using probability theory. We record a deterministic example for disproving this estimate.

1. Introduction

The question of determining the validity of the following statement was raised in [**2**, Remark 1] and was recently answered negatively in [**6**]

(W) $\quad \|e^{it|D|}u\|_{L^2_t L^\infty_x} \lesssim \|u\|_{\dot{H}^{\frac{n-1}{2}}} \quad$ holds for all $u \in \mathcal{S}(\mathbb{R}^n)$ when $n \geq 4$.

The function $v = e^{it|D|}u$ solves the wave equation $(-\partial_t^2 + \Delta)v = 0$. Here $\mathcal{S}(\mathbb{R}^n)$ denotes the space of Schwartz functions, while $\dot{H}^s$ denote homogeneous Sobolev spaces.

In fact [**6**] disproved estimates $e^{it|D|^a} : \dot{H}^{\frac{n-a}{2}} \to L^2_t L^\infty_x$ for general operators given by $a \in (0, 2]$, using stable Levy processes. The case $a = 2$ corresponds to the Schrödinger equation.

In this note we record a proof of the failure of (W) using basic Fourier analysis. This argument is specific to the wave equation and does not apply to the Schrödinger case.

Theorem 1.1. *The statement* (W) *is not true.*

Estimates of $L^q_t L^r_x$ norms of solutions to dispersive equations with initial conditions in $\dot{H}^s$ spaces are known as Strichartz estimates, see [**20**], [**24**], [**7**], [**4**], [**12**], [**5**], [**17**], [**23**]. The exponents $(q, r) = (2, \bar{r})$ when r is finite and minimal[1] (blue point in Figure 1 Left) are called the endpoints, and have been famously resolved in [**8**]. We refer to [**23**, Section 2.3, see "Knapp example"] for a discussion of the minimality of $\bar{r}$.

The case $r = \infty$ is usually omitted when stating Strichartz estimates in higher dimensions ($n \geq 4$ for the Wave equation and $n \geq 3$ for the Schrödinger equation), see for example [**23**], [**10**], [**17**], because the strongest and most useful cases occur on the blue line in Fig. 1, i.e. the range between $L^2_t L^{\bar{r}}_x$ and $L^\infty_t L^2_x$. Other cases

2020 *Mathematics Subject Classification.* Primary 35L05.

[1] i.e. $\bar{r} = \frac{2(n-1)}{n-3}$ for the Wave equation ($n \geq 4$) and $\bar{r} = \frac{2n}{n-2}$ for the Schrödinger eq. ($n \geq 3$).

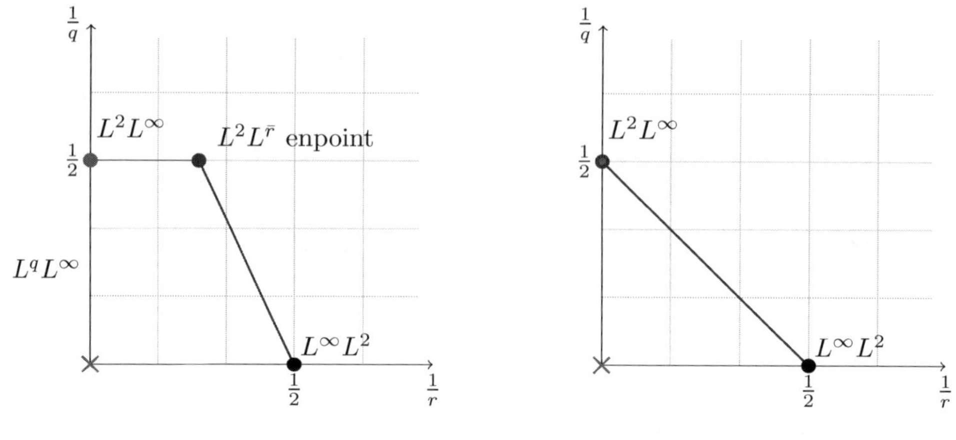

FIGURE 1. Left: high dimensions. Right: $n = 3$ for W.E.

are typically obtained by Sobolev embeddings. The $L_t^2 L_x^\infty$ exponents differ from the endpoints, unless $n = 3$ for the Wave equation or $n = 2$ for the Schrödinger equation, Figure 1 Right. In those low-dimension cases, counterexamples are given in [**9**], [**15**]. See Remark 1.2 below. The present case is the red point in Fig. 1 Left.

The case $r = \infty$ is generally delicate because $\dot{W}^{\frac{n}{r},r} \not\subseteq L^\infty$, i.e. Sobolev embeddings fail into L^∞. Thus (W) does not follow from the Keel-Tao endpoint $L_t^2 L_x^{\bar{r}}$ in [**8**], and instead one can only conclude from [**8**] estimates that are true if one replaces L_x^∞ with BMO or Besov spaces (or $L_t^2 L_x^\infty$ estimates restricted to frequency localized functions) - see the ending of Remark 1.2.

Estimates $e^{it|D|} : \dot{H}^{\frac{n}{2}-\frac{1}{q}} \to L_t^q L_x^\infty$ for $q \in (2, \infty)$ are proved in [**11**], [**2**]. The case $q = \infty$ trivially fails at $t = 0$ due to $\dot{H}^{\frac{n}{2}} \not\subseteq L_x^\infty$. The case that remained was $q = 2$, which was recently treated in [**6**].

REMARK 1.2. When $n = 3$ the estimate in (W) is disproved in [**9**, Remark 1] (also presented in [**10**]), exhibiting concentration along null rays. Using the $3d$ fundamental solution (i.e. Kirchhoff's formula) they show there exists a sequence of solutions to $(-\partial_t^2 + \Delta)\phi_n = 0$ with normalized initial velocities $\|\partial_t \phi_n(0, \cdot)\|_{L^2} = 1$ and $\phi_n(0, \cdot) = 0$ such that

$$\int_0^\infty |\phi_n(t, te_1)|^2 \, \mathrm{d}t \to \infty \qquad \text{as } n \to \infty.$$

See also [**21**, Prop. 1.1] for an explicit example of such initial velocity for which the solution has $\phi(t, te_1) = \infty$ for all $t \in (1, 2)$ (presented as an adaptation of Stein's counterexample in regards to the spherical maximal function).

Moreover, [**15**] shows that, when $n = 3$, (W) fails even with L_x^∞ replaced by BMO. This is in contrast to the case $n \geq 4$ for which that variant can be obtained from the Keel-Tao endpoint [**8**] and the embedding $\dot{W}^{\frac{n}{r},r} \subset BMO$.

The proof of Theorem 1.1 proceeds instead in Fourier space and must use rays of velocity > 1 instead of lightrays.

1.1. Notation. In what follows $\hat{f}$ denotes the Fourier transform of f. Operators $m(D)$, such as $e^{it|D|}, e^{it\Delta}$, are defined by $\widehat{m(D)f}(\xi) = m(\xi)\hat{f}(\xi)$. Here

$\Delta = -|D|^2$. Homogeneous Sobolev spaces are defined by $\dot{H}^s = |D|^{-s} L^2(\mathbb{R}^n)$ for $s < \frac{n}{2}$. Space-time norms are defined by

$$\|F\|_{L_t^q L_x^r}^q = \int_{-\infty}^{\infty} \|F(t,\cdot)\|_{L_x^r(\mathbb{R}^n)}^q \,\mathrm{d}t.$$

Littlewood-Paley projections $P_{\leq k}$ are defined by $\widehat{P_{\leq k}h}(\xi) = \chi(\frac{|\xi|}{2^k})\hat{h}(\xi)$, where $\chi \in C_c^\infty(\mathbb{R})$, $\chi \geq 0$ and $\chi(\eta) = 1$ for η near 0. For example, $\widehat{P_{\leq k}\delta_0}(\xi) = \chi(\frac{|\xi|}{2^k})$. Here δ_0 denotes the Dirac delta measure. $S^{n-1} \subset \mathbb{R}^n$ denotes the unit sphere. The notation $X \lesssim Y$ means $X \leq CY$ for an absolute constant $0 < C < \infty$. When $Y \lesssim X \lesssim Y$ we denote $X \simeq Y$.

2. Proof of Theorem 1.1

Assume that (W) is true. Then, by duality, one also has

$$\text{(W')} \qquad \left\| \int_{-\infty}^{\infty} e^{-it|D|} \frac{1}{|D|^{\frac{n-1}{2}}} F(t)\,\mathrm{d}t \right\|_{L_x^2} \lesssim \|F\|_{L_t^2 L_x^1}$$

for all $F \in L_t^2 L_x^1 \cap L_t^1 \dot{H}_x^{-\frac{n-1}{2}}$.

Take any $f \in L_t^1 \cap L_t^2$ with $\|f\|_{L^2} = 1$ such that $\int_0^\infty |\hat{f}(\eta)|^2 \,\mathrm{d}\eta \neq 0$. Consider also the function $h \in L_x^1 \cap \dot{H}_x^{-\frac{n-1}{2}}$ with $\|h\|_{L_x^1} = 1$ which remains to be defined. Fix a direction $\omega \in S^{n-1}$ and a constant $c > 0$. Applying (W') with

$$F(t,x) := f(t)h(x - ct\omega)$$

and using Plancherel's theorem one obtains

$$\left\| \frac{1}{|\xi|^{\frac{n-1}{2}}} \hat{h}(\xi) \int_{-\infty}^{\infty} e^{-it|\xi|} e^{-ict\omega\cdot\xi} f(t)\,\mathrm{d}t \right\|_{L_\xi^2} \lesssim 1$$

which implies

$$\left\| |\xi|^{-\frac{n-1}{2}} \hat{h}(\xi) \hat{f}(|\xi| + c\omega\cdot\xi) \right\|_{L_\xi^2} \lesssim 1.$$

Consider $h = \delta_0$, $\hat{h} \equiv 1$, or rather, since $\delta_0 \notin L^1 \cap \dot{H}^{-\frac{n-1}{2}}$, take an approximation to the identity sequence $h_k = P_{\leq k}\delta_0$ and pass to the limit. Here δ_0 denotes the Dirac delta measure. By the Monotone Convergence theorem one obtains

$$\int_{\mathbb{R}^n} \frac{1}{|\xi|^{n-1}} |\hat{f}(|\xi| + c\omega\cdot\xi)|^2 \,\mathrm{d}\xi \lesssim 1.$$

Write this integral in spherical coordinates, where

$$|\xi| = \lambda, \quad \omega\cdot\xi = \lambda\cos\theta, \quad \theta \in [0,\pi].$$

This implies

$$\int_0^\infty \int_0^\pi |\hat{f}(\lambda(1 + c\cos\theta))|^2 (\sin\theta)^{n-2} \,\mathrm{d}\theta \,\mathrm{d}\lambda \lesssim 1.$$

REMARK 2.1. For $n = 3$ one may take $c = 1$ and change coordinates to $u = 1 + \cos\theta$ to obtain

$$\int_0^\infty \int_0^2 |\hat{f}(\lambda u)|^2 \,\mathrm{d}u \,\mathrm{d}\lambda = \int_0^2 \frac{1}{u} \,\mathrm{d}u \int_0^\infty |\hat{f}(\eta)|^2 \,\mathrm{d}\eta \lesssim 1$$

which gives a contradiction. One may view this as a Fourier version of the physical space construction in Remark 1.2 above, due to [**9**]. Δ

For $n \geq 4$, choosing $c = 1$ does not work. Instead, take $c = 2$ and restrict the integral to $\theta \in [\frac{\pi}{2}, \frac{2\pi}{3}]$. On this interval $\sin\theta \simeq 1$. As before, one obtains

$$\int_0^\infty |\hat{f}(\eta)|^2 \,\mathrm{d}\eta \int_{\frac{\pi}{2}}^{\frac{2\pi}{3}} \frac{(\sin\theta)^{n-2}}{1+2\cos\theta} \,\mathrm{d}\theta \lesssim 1.$$

The last integral is logarithmically divergent, being $\simeq \int_0^1 \frac{1}{y}\,\mathrm{d}y$, thus obtaining a contradiction again, which concludes the proof. $\square$

References

[1] Philip Brenner, *On $L_p - L_{p'}$ estimates for the wave-equation*, Math. Z. **145** (1975), no. 3, 251–254, DOI 10.1007/BF01215290. MR387819

[2] Daoyuan Fang and Chengbo Wang, *Some remarks on Strichartz estimates for homogeneous wave equation*, Nonlinear Anal. **65** (2006), no. 3, 697–706, DOI 10.1016/j.na.2005.09.040. MR2231083

[3] Damiano Foschi, *Inhomogeneous Strichartz estimates*, J. Hyperbolic Differ. Equ. **2** (2005), no. 1, 1–24, DOI 10.1142/S0219891605000361. MR2134950

[4] J. Ginibre and G. Velo, *Smoothing properties and retarded estimates for some dispersive evolution equations*, Comm. Math. Phys. **144** (1992), no. 1, 163–188. MR1151250

[5] J. Ginibre and G. Velo, *Generalized Strichartz inequalities for the wave equation*, J. Funct. Anal. **133** (1995), no. 1, 50–68, DOI 10.1006/jfan.1995.1119. MR1351643

[6] Zihua Guo, Ji Li, Kenji Nakanishi, and Lixin Yan, *On the boundary Strichartz estimates for wave and Schrödinger equations*, J. Differential Equations **265** (2018), no. 11, 5656–5675, DOI 10.1016/j.jde.2018.07.010. MR3857493

[7] L. V. Kapitanskiĭ, *Some generalizations of the Strichartz-Brenner inequality* (Russian), Algebra i Analiz **1** (1989), no. 3, 127–159; English transl., Leningrad Math. J. **1** (1990), no. 3, 693–726. MR1015129

[8] Markus Keel and Terence Tao, *Endpoint Strichartz estimates*, Amer. J. Math. **120** (1998), no. 5, 955–980. MR1646048

[9] S. Klainerman and M. Machedon, *Space-time estimates for null forms and the local existence theorem*, Comm. Pure Appl. Math. **46** (1993), no. 9, 1221–1268, DOI 10.1002/cpa.3160460902. MR1231427

[10] Sergiu Klainerman, *Lecture notes in analysis (2011)*, `https://web.math.princeton.edu/~seri/homepage/courses/Analysis2011.pdf`.

[11] Sergiu Klainerman and Matei Machedon, *On the algebraic properties of the $H_{n/2,1/2}$ spaces*, Internat. Math. Res. Notices **15** (1998), 765–774, DOI 10.1155/S1073792898000464. MR1639547

[12] Hans Lindblad and Christopher D. Sogge, *On existence and scattering with minimal regularity for semilinear wave equations*, J. Funct. Anal. **130** (1995), no. 2, 357–426, DOI 10.1006/jfan.1995.1075. MR1335386

[13] Shuji Machihara, Makoto Nakamura, Kenji Nakanishi, and Tohru Ozawa, *Endpoint Strichartz estimates and global solutions for the nonlinear Dirac equation*, J. Funct. Anal. **219** (2005), no. 1, 1–20, DOI 10.1016/j.jfa.2004.07.005. MR2108356

[14] Gerd Mockenhaupt, Andreas Seeger, and Christopher D. Sogge, *Local smoothing of Fourier integral operators and Carleson-Sjölin estimates*, J. Amer. Math. Soc. **6** (1993), no. 1, 65–130, DOI 10.2307/2152795. MR1168960

[15] S. J. Montgomery-Smith, *Time decay for the bounded mean oscillation of solutions of the Schrödinger and wave equations*, Duke Math. J. **91** (1998), no. 2, 393–408, DOI 10.1215/S0012-7094-98-09117-7. MR1600602

[16] Irving Segal, *Space-time decay for solutions of wave equations*, Advances in Math. **22** (1976), no. 3, 305–311, DOI 10.1016/0001-8708(76)90097-9. MR492892

[17] Christopher D. Sogge, *Lectures on nonlinear wave equations*, Monographs in Analysis, II, International Press, Boston, MA, 1995. MR1715192

[18] Jacob Sterbenz, *Angular regularity and Strichartz estimates for the wave equation*, Int. Math. Res. Not. **4** (2005), 187–231, DOI 10.1155/IMRN.2005.187. With an appendix by Igor Rodnianski. MR2128434

[19] Robert S. Strichartz, *A priori estimates for the wave equation and some applications*, J. Functional Analysis **5** (1970), 218–235, DOI 10.1016/0022-1236(70)90027-3. MR0257581

[20] Robert S. Strichartz, *Restrictions of Fourier transforms to quadratic surfaces and decay of solutions of wave equations*, Duke Math. J. **44** (1977), no. 3, 705–714. MR512086

[21] Terence Tao, *Counterexamples to the n=3 endpoint strichartz estimate for the wave equation*, Expository note: `https://www.math.ucla.edu/~tao/preprints/Expository/stein.dvi`.

[22] Terence Tao, *A counterexample to an endpoint bilinear Strichartz inequality*, Electron. J. Differential Equations (2006), No. 151, 6. MR2276576

[23] Terence Tao, *Nonlinear dispersive equations*, CBMS Regional Conference Series in Mathematics, vol. 106, Published for the Conference Board of the Mathematical Sciences, Washington, DC; by the American Mathematical Society, Providence, RI, 2006. Local and global analysis, DOI 10.1090/cbms/106. MR2233925

[24] Kenji Yajima, *Existence of solutions for Schrödinger evolution equations*, Comm. Math. Phys. **110** (1987), no. 3, 415–426. MR891945

Department of Mathematics, Johns Hopkins University, Baltimore, Maryland 21218
Email address: `cgavrus1@jhu.edu, gcd2006@gmail.com`

Contemporary Mathematics
Volume **781**, 2023
https://doi.org/10.1090/conm/781/15707

On the Liouville property for fully nonlinear equations with superlinear first-order terms

Marco Cirant and Alessandro Goffi

ABSTRACT. We consider in this note one-side Liouville properties for viscosity solutions of various fully nonlinear uniformly elliptic inequalities, whose prototype is $F(x, D^2u) \geq H_i(x, u, Du)$ in $\mathbb{R}^N$, where H_i has superlinear growth in the gradient variable. After a brief survey on the existing literature, we discuss the validity or the failure of the Liouville property in the model cases $H_1(u, Du) = u^q + |Du|^\gamma$, $H_2(u, Du) = u^q|Du|^\gamma$ and $H_3(x, u, Du) = \pm u^q|Du|^\gamma - b(x) \cdot Du$, where $q \geq 0$, $\gamma > 1$ and b is a suitable velocity field. Several counterexamples and open problems are thoroughly discussed.

1. Introduction

The question of non-existence of non-trivial solutions to linear and nonlinear partial differential equations and inequalities satisfying one or two-side bounds - usually called *Liouville-type property* - has been widely investigated in the literature. This qualitative property is crucial in many theoretical aspects of PDEs, ranging from regularity theory to quantitative properties, see e.g. the monographs [**76, 81**].

The Liouville property for solutions to PDEs is usually obtained as a consequence of the (invariant) Harnack inequality, see e.g. [**24, 42, 48, 54**]. Other ways to deduce such Liouville-type results are based on mean-value formulas [**22, 67**] or on a priori gradient estimates, that can be derived either via maximum principle methods, as in [**72**], or via Bernstein-type methods, as started in [**68, 79**], both for elliptic and parabolic equations, see also the recent work [**27**] along with the discussion and references in [**30**]. We refer also to [**55**] for a quite different approach based on a comparison principle on punctured domains and to [**73**] for probabilistic methods as well as for a discussion on control theoretic interpretations.

The case of solutions to partial differential inequalities satisfying a one-side bound is somehow different. It is well-known that superharmonic (subharmonic)

2020 *Mathematics Subject Classification.* Primary 35J60, 35B53, 35D40.

Key words and phrases. Fully nonlinear equation, Liouville theorems, Hamilton-Jacobi equations, Lane-Emden equations, superlinear gradient terms.

The authors are members of the Gruppo Nazionale per l'Analisi Matematica, la Probabilità e le loro Applicazioni (GNAMPA) of the Istituto Nazionale di Alta Matematica (INdAM). This research was carried out while the second-named author was Postdoctoral researcher at the University of Padova. The authors warmly thank the anonymous referees for their careful reading and the suggestions that improved the presentation of the first draft of the manuscript. The authors wish to thank Prof. Roberta Filippucci and Dr. Giulio Tralli for fruitful discussions and for suggesting many references on the matter of this paper.

functions bounded from below (above) in $\mathbb{R}^N$ are constants provided that $N \leq 2$. This result can be obtained via the Hadamard three-circle theorem [**74**, Theorem 29 Ch. 2] or via different methods involving the existence of a so-called Lyapunov function (alternatively known as Khas'minskii test, cf [**59**] and the references therein), see [**9**, Theorem 2.1 and Remark 2.2], [**7**, Theorem 2.1] and also [**71**, p. 230-(ii)]. The underlying motive is the unboundedness of the fundamental solution of the Laplace equation. This result is sharp as there exist counterexamples in higher dimensions, see the introduction in [**9**]. Nonetheless, it is also by now well-known that the perturbation of second-order operators with suitable nonlinearities, such as zero-th and first-order terms, allows to recover the Liouville property in higher dimensions.

The first analysis on supersolutions to semi-linear equations started with B. Gidas [**40**]. When the equation is driven by a linear (or even quasi-linear) operator, the typical strategies to derive Liouville properties for inequalities with zero-th and/or first order terms are two. The first one was developed during the 90's by M.-F. Bidaut-Véron [**16**], H. Berestycki- I. Capuzzo-Dolcetta- L. Nirenberg [**12**], E. Mitidieri-S. Pohozaev [**33, 62, 63**] and V. Kurta [**52, 53**], see also the references therein, and it is based on integral estimates and test function methods via the weak formulation of the problem. See also the references given in [**15**] for recent developments. Moreover, such integral methods provide finer results that are known in the literature as Liouville comparison principles, as studied e.g. by V. Kurta and V. Kurta-B. Kawohl, see for instance [**47, 52, 53**]. When the Liouville property fails, the latter analysis allows even to establish the sharp distance at infinity of the non-constant supersolution bounded below by a constant to the constant itself, cf e.g. [**53**, Theorem 1.2] and the references therein. The second approach basically consists in "radializing" the equation: when it is driven by the Laplacian, one can observe that the spherical mean of the unknown function satisfies again a partial differential inequality, reducing the analysis to an ODE problem, cf [**26**, Section 7]. We refer to the introduction of [**9**] and to [**15, 46**] for further updated references on the quasi-linear and semi-linear (even degenerate) case for problems posed on Riemannian manifolds and Carnot groups.

However, the aforementioned methods break down in the fully nonlinear setting, and actually fail even for linear problems driven by operators in non-divergence form with non-differentiable diffusion matrix and/or having unbounded drifts, such as those of Ornstein-Uhlenbeck type. To the authors' knowledge, there are few established approaches to handle one-side Liouville properties for fully nonlinear partial differential inequalities, and all of them are obviously based on maximum principle arguments, as the right setting to investigate such equations is that of viscosity solutions [**24, 31**].

The first one was initiated by A. Cutrì and F. Leoni [**32**]. They proved non-existence of non-trivial solutions (in the viscosity sense) to $F(x, D^2u) \geq u^q$ in $\mathbb{R}^N$ (or more generally to problems driven by rotationally invariant operators perturbed with a weighted radial zero-th order term) when q is below a certain exponent related to the scaling of the equation, via a nonlinear Hadamard three-sphere theorem. This approach is based on the following observation: any nonnegative solution to $F(x, D^2u) \geq u^q$ in $\mathbb{R}^N$ is also a solution to $\mathcal{P}^+_{\lambda,\Lambda}(D^2u) \geq 0$ in $\mathbb{R}^N$, where $\mathcal{P}^+_{\lambda,\Lambda}$ stands for the Pucci's maximal operator (see below for the definition) and it satisfies by the maximum principle $\min_{\partial B(0,r)} u = \min_{\bar{B}(0,r)} u$, which implies that the function

$r \longmapsto m(r) = \min_{\partial B(0,r)} u$ is decreasing. Then, owing to the radial symmetry of the fundamental solution of the extremal operators, one can establish a nonlinear Hadamard three-sphere theorem, and then deduce the crucial fact that the function $r^{\beta-2}m(r)$ is increasing, $\beta - 2$ being the scaling exponent of the Pucci's maximal operator (which will be specified below), see [**32**, Corollary 3.1]. Finally, the interplay between the comparison principle, the strong minimum principle and the construction of a suitable test function allows to prove that $m(r)$ approaches zero as $r \to \infty$ with a certain rate, contradicting the previous step. Alternatively, the monotonicity of the function $r^{\beta-2}m(r)$ can be combined with the weak Harnack inequality from [**24**] to get the same conclusion, cf e.g. [**32**, Remark 9]. The approach via the Hadamard three-sphere theorem was later refined in [**25**] for problems with gradient terms having at most linear growth, under appropriate smallness conditions on the drift coefficient. See also [**35**] for Liouville theorems under Keller-Osserman type conditions. Other related results based on qualitative properties of the function $m(r)$ have been obtained in [**1, 2, 19, 20**], see also [**21, 43**] for further refinements for degenerate problems.

Other results for fully nonlinear equations having at most linear growth in the gradient have been obtained in [**77**], under suitable conditions at infinity on the coefficients. The approach is based on a strong maximum principle and the construction of suitable test functions. We refer to the references given in [**77**] for further results on semi-linear and quasi-linear equations under similar conditions on the coefficients.

More general approaches to handle Bellman-Isaacs equations (even those driven by the p-Laplacian and/or set on exterior domains) have been developed in [**4, 5**] by S. Armstrong and B. Sirakov. The results obtained in [**5**] are based again on maximum principles arguments and the analysis of "fundamental solutions" to such operators initiated in [**6**]. These somehow extended [**32**, Theorem 4.1] to a wider framework when the equation is perturbed by zero-th order terms. Instead, the results in [**4**] are based on a quantitative version of the strong maximum principle, cf [**4**, Theorem 3.3], and the method of proof allows to encompass both equations driven by Bellman-Isaacs and p-Laplacian operators. We emphasize that the dividing exponent implying the Liouville property in the general case of a Bellman-Isaacs operator is not explicit as in the case of rotationally invariant operators, as it depends on the intrinsic dimension of its fundamental solution.

It is well-known that the Liouville property for linear operators is strictly related to the recurrence of the associated diffusion process, which is in turn tied up with the existence of a so-called *Lyapunov function*, see for example [**45, 60**] and references therein. The argument showing that the Liouville property follows from the existence of Lyapunov functions (or the validity of the Khas'minskii test) is particularly flexible: it basically needs the strong maximum/minimum principle and the comparison principle on bounded open sets, and does not go through more sophisticated steps such as regularity properties or Harnack inequalities (as [**4, 6**]). Maximum and comparison principles are two ingredients that are broadly available for elliptic PDEs (cf [**8, 44**] for recent developments in the viscosity solutions' framework). This strategy has been consequently extended to the nonlinear framework in the context of Riemannian manifolds on one hand, see [**15, 59, 69–71**], and, on the other hand, in [**7**] and later in [**9, 10**], with focus on degenerate problems, possibly modeled over Hörmander vector fields. We refer also to [**57**] and [**7**, **50**, **58**] for related Liouville properties and applications to ergodic control.

We remark in passing that the non-existence properties in the fully nonlinear setting transfer immediately to non-divergence equations (i.e. involving the operator $-\mathrm{Tr}(A(x)D^2u)$), being the Pucci's extremal operators the prototype ones in non-divergence form, see Remark 5.1. We emphasize that the literature on Liouville properties driven by non-divergence operators is poor to our knowledge, see e.g. [**51**] for some results in this direction for semi-linear equations.

Finally, we mention that different Liouville properties for partial differential equations and inequalities have been proved for solutions belonging to L^p scales: for this topic, without being exhaustive, we refer to [**45**, Section 13.1], [**49**], [**71**] and the references therein.

In this note, our main motivations in Liouville-type results are their implications in regularity properties and existence of solutions to nonlinear elliptic equations [**29**, **41**, **78**] (that can be established via blow-up procedures), their qualitative behavior [**14**, **81**], ergodic control problems and the large time behavior for nonlinear parabolic equations, cf e.g. [**7**], and the recent developments in the theory of Mean Field Games [**28**], where systems of equations with gradient dependent nonlinearities appear naturally. In particular, we aim at investigating Liouville properties for supersolutions bounded below of fully nonlinear equations, with particular emphasis on problems involving a first-order term with superlinear growth in the gradient. We will consider the following inequality in the viscosity sense,

$$F(x, D^2u) \geq H_i(x, u, Du) \text{ in } \mathbb{R}^N,$$

where F is uniformly elliptic, i.e.

$$\mathcal{P}^-_{\lambda,\Lambda}(M-Q) \leq F(x,M) - F(x,Q) \leq \mathcal{P}^+_{\lambda,\Lambda}(M-Q)\ , M, Q \in \mathrm{Sym}_N\ , Q \geq 0\ ,$$

where Sym_N stands for the space of $N \times N$ symmetric matrices. Here, $\mathcal{P}^\pm_{\lambda,\Lambda}$ are the Pucci's extremal operators [**24**] defined for ellipticity constants $0 < \lambda \leq \Lambda$ as

$$\mathcal{P}^+_{\lambda,\Lambda}(M) = \sup_{\lambda I_N \leq A \leq \Lambda I_N} -\mathrm{Tr}(AM) = -\lambda \sum_{e_k>0} e_k - \Lambda \sum_{e_k<0} e_k\ ,$$

$$\mathcal{P}^-_{\lambda,\Lambda}(M) = \inf_{\lambda I_N \leq A \leq \Lambda I_N} -\mathrm{Tr}(AM) = -\Lambda \sum_{e_k>0} e_k - \lambda \sum_{e_k<0} e_k\ ,$$

where e_k stands for the k-th eigenvalue of the matrix M. We will further assume the normalization condition $F(x, 0) = 0$, and discuss the validity of the Liouville property for the following model nonlinearities

$$H_1(r,\xi) = r^q + |\xi|^\gamma\ ,$$
$$H_2(r,\xi) = r^q|\xi|^\gamma\ ,$$
$$H_3(x,r,\xi) = \pm r^q|\xi|^\gamma - b(x)\cdot\xi\ ,$$

with $q \geq 0$, $\gamma > 1$, and b being a suitable confining velocity field. In the first two cases it clearly holds that $\mathcal{P}^+_{\lambda,\Lambda}(D^2u) \geq 0$, and this fact will be frequently exploited.

We now recall some related results for problems driven by linear operators. First, it is worth recalling that when $H = H_2$ and $\lambda = \Lambda = 1$, in the case $q = 0$ the equation reduces to the so-called *viscous Hamilton-Jacobi equation* [**29**, **56**], while when $\gamma = 0$ it reduces to the celebrated *Lane-Emden equation* [**40**] (see [**5**, Theorem

1.1] for results in this direction). For $q \geq 0$, $\gamma > 1$, it has been first proved in [**26**] that nonnegative classical solutions to

$$-\Delta u \geq u^q |Du|^\gamma \text{ in } \mathbb{R}^N \ , N > 2 \ ,$$

must be constants provided that

$$(N-2)q + (N-1)\gamma \leq N \ .$$

The proof in [**26**, Theorem 7.1] is based on replacing u by its spherical mean, which turns out to be a supersolution by means of Jensen's inequality. This allows to reduce the study of the Liouville property to radial supersolutions, hence to an ODE analysis. Instead, a different study was performed in e.g. [**47**] and [**63**, Theorem 15.1] via the nonlinear capacity method combined with an asymptotic analysis, see also [**23**, Corollary 3], the recent papers [**18**, Theorem 2.1], [**27**, Theorem 1.3], and the earlier results in [**37**, Corollary 1] together with the further developments obtained therein for elliptic systems of inequalities with weights, even driven by p-Laplacians. We finally mention the recent paper [**80**] which studies the elliptic inequality $-\Delta_p u \geq u^q |Du|^\gamma$ on Riemannian manifolds under volume growth conditions, even for negative values of q, γ. Note that these latter results for negative pairs are new even in the Euclidean case. We remark that [**36**, Corollary 1.1] provides the result when $\gamma < 1$ and for operators driven by the p-Laplacian, see [**20**] and [**23**] for a different proof based on the maximum principle. In the fully nonlinear case, the only available Liouville property for the inequality $F \geq H_2$ in $\mathbb{R}^N$ was obtained in [**20**] when $q > 0$ and $0 < \gamma < 1$, see Remark 3.6. The case $H = H_1$ driven by the Laplacian in exterior domains has been first tackled in [**26**, Theorem 7.2] again via the radialization through the spherical mean. The same problem has been recently investigated in much more generality in [**1**] via the methods initiated in [**32**], where the Liouville property for

$$-\Delta u \geq u^q + |Du|^\gamma \quad \text{ in } \mathbb{R}^N \backslash B(0, R_0)$$

has been proved when $q > \frac{N}{N-2}$ and $\gamma \leq \frac{N}{N-1}$ (the case $q \leq \frac{N}{N-2}$ and $\gamma > 0$ being a trivial consequence of the result for supersolutions to $-\Delta u \geq u^q$). This result is sharp, since one can find counterexamples to the Liouville property when $\gamma > \frac{N}{N-1}$, see Remark 2.5 for an explicit calculation.

The case $H = H_3$ with $q = \gamma = 0$ and $F = \mathcal{P}^+$ has been first tackled in [**7**] in the Euclidean case and in [**9**] for subelliptic problems. In particular, assuming the existence of a sort of Lyapunov function, namely of a viscosity subsolution w outside a compact set such that $\lim_{|x|\to\infty} w(x) = -\infty$, it has been proved that if $u \in \mathrm{LSC}$ is a supersolution to

$$\mathcal{P}^+_{\lambda,\Lambda}(D^2 u) + \sup_{\alpha\in A}\{c^\alpha(x)u - b^\alpha(x)\cdot Du\} = 0 \text{ in } \mathbb{R}^N$$

satisfying $\limsup_{|x|\to\infty} \frac{u(x)}{w(x)} \leq 0$, then u is constant. The existence of such w is shown, for example, whenever

$$\sup_{\alpha\in A}\{b^\alpha(x)\cdot x - c^\alpha(x)|x|^2 \log|x|\} \leq \lambda - \Lambda(N-1)$$

for large $|x|$.

In this note we will discuss the following results, which depend on the *effective dimension*

$$\beta = \frac{\Lambda}{\lambda}(N-1) + 1$$

of the Pucci's maximal operator, which is in turn related to the scaling exponents of the nonlinearities.

- When $H = H_1$ and F is uniformly elliptic, we prove the validity of the Liouville property when $N \geq 3$, and either $q \leq \frac{\beta}{\beta-2}$ and $\gamma > 1$ (actually $\gamma > 0$ is allowed), or $q > \frac{\beta}{\beta-2}$ and $1 < \gamma \leq \frac{\beta}{\beta-1}$, see Theorem 2.1. We further prove the sharpness of the result when the above growth conditions fail, see Remark 2.5. To our knowledge, these properties are new, although the proof combines known ideas from [**1, 32**]. Indeed, nonnegative solutions to $F \geq H_1$ are also solutions to $\mathcal{P}^+_{\lambda,\Lambda}(D^2u) \geq u^q$ in $\mathbb{R}^N$, a fact that is crucial to establish the Liouville property;
- When $H = H_2$, we prove that the Liouville property fails when
$$(\beta - 2)q + (\beta - 1)\gamma > \beta \;, q \geq 0 \;, \gamma > 1.$$
In particular, when $q = 0$ we prove in Theorem 3.1 that for a suitable constant $c > 0$ the function $u(x) = c(1+|x|^2)^{-\frac{2-\gamma}{2(\gamma-1)}}$ is a classical solution to $\mathcal{P}^+_{\lambda,\Lambda}(D^2u) \geq |Du|^\gamma$ in $\mathbb{R}^N$, provided that
$$\gamma > \frac{\beta}{\beta - 1} \;.$$
Based on this observation, we conjecture its validity when
$$(\beta - 2)q + (\beta - 1)\gamma \leq \beta, q \geq 0 \;, \gamma > 1,$$
and in particular that for the viscosity inequality $F(x, D^2u) \geq |Du|^\gamma$ in $\mathbb{R}^N$ the Liouville property holds provided that $1 < \gamma \leq \frac{\beta}{\beta-1}$. This problem will be addressed in a future research.
- When $H = H_3$ and $q = 0$, we provide a sufficient condition to the Liouville property involving the behavior of the drift b , for problems with arbitrary "repulsive" gradient terms, or quadratic "absorbing" terms (via a nonlinear Hopf-Cole transformation), see Theorem 4.1. These properties are new, and will be derived from the results in [**7, 9**], which use Lyapunov functions (Khas'minskii test). We will show in Theorem 4.7 that such condition can be improved in some special cases (such as for supersolutions to equations driven by $F = \mathcal{P}^-$). Moreover, a variant of the Hopf-Cole transformation for fully nonlinear equations allows to treat the case $H(u, Du) = u^q|Du|^2$, see Remark 4.9. For these drift perturbed equations, we show that the condition ensuring the Liouville property is sharp, see Remark 4.6. Therefore, we prove that our nonlinear version of the Khas'minskii test is not only sufficient, but also necessary for the validity of the Liouville property, at least in the two cases $F = \mathcal{P}^\pm$.
- In Section 5 we make some remarks on degenerate equations, on the cases $N = 1, 2$ for H_i, $i = 1, 2, 3$, and on some problems driven by non-divergent operators. Finally, we make in Section 6 some remarks on the Liouville property for solutions to Hamilton-Jacobi equations. Some open problems will also be mentioned throughout the paper.

Finally, let us stress that some of our results are new even for supersolutions to the equation
$$-\mathrm{Tr}(A(x)D^2u) = H_i(x, u, Du) \text{ in } \mathbb{R}^N \;,$$

with $\lambda I_N \leq A \leq \Lambda I_N$, cf Remark 5.1. We defer to a future study the proof of the Liouville property for entire solutions to

$$F(x, D^2u) \geq u^q |Du|^\gamma \text{ in } \mathbb{R}^N$$

as well as for a general treatment of the aforementioned properties for Bellman-Isaacs equations.

2. Liouville properties for fully nonlinear equations involving the sum of first- and zero-th order terms with power-like growth

In this section we consider Liouville properties for solutions to viscosity inequalities in the whole space such as

$$F(x, D^2u) \geq u^q + |Du|^\gamma \quad \text{in } \mathbb{R}^N , \tag{2.1}$$

where $F : \mathbb{R}^N \times \mathrm{Sym}_N \to \mathbb{R}$ is continuous, uniformly elliptic and satisfies the normalization condition $F(x, 0) = 0$, while $q > 0$, $\gamma > 1$. First, by the definition of uniform ellipticity, one observes that viscosity solutions to (2.1) are also solutions to the viscosity inequality

$$\mathcal{P}^+_{\lambda,\Lambda}(D^2u) \geq u^q \text{ in } \mathbb{R}^N .$$

Then, [**5**, Theorem 1.4-(i)] or [**32**, Theorem 4.1] ensure that (2.1) has no nontrivial nonnegative supersolutions in $\mathbb{R}^N$ when $0 < q \leq \frac{\beta}{\beta-2}$, where

$$\beta = \frac{\Lambda}{\lambda}(N-1) + 1 .$$

Such a threshold is sharp, since for $q > \frac{\beta}{\beta-2}$ the function $u(x) = c(1 + |x|^2)^{-\frac{1}{q-1}}$ satisfies in the above inequality for a suitable $c > 0$. Therefore, we restrict our attention to the case $q > \frac{\beta}{\beta-2}$, and add the nonlinearity $|Du|^\gamma$ to the problem. We characterize below the (sharp) interval for γ that guarantees the Liouville property.

The next result shows indeed that the presence of a (slightly) superlinear first order term allows to extend the range of the validity of the Liouville property even beyond the threshold $q = \frac{\beta}{\beta-2}$.

THEOREM 2.1. *Assume that $q > \frac{\beta}{\beta-2}$ and $1 < \gamma \leq \frac{\beta}{\beta-1}$. Then, any nonnegative viscosity supersolution to* (2.1) *is constant.*

The proof of Theorem 2.1 will be based on the analysis of

$$m(R) := \min_{|x|\leq R} u(x)$$

, where u is the supersolution to (2.1). for $R \geq 0$, which combines the strong minimum principle, see e.g. [**8, 44**], and the comparison principle [**31**]. We will follow some ideas developed in [**32**] and [**1**]. Note that by the strong minimum principle we can always assume $u > 0$ in $\mathbb{R}^N$, so that $m(R) > 0$. Indeed, if there existed a point $x_0 \in B(0, R)$ such that u attains its minimum at x_0, it would follow $u \equiv 0$. Then, by the arbitrariness of $R > 0$ we would have $u \equiv 0$ in $\mathbb{R}^N$.

We start with the following algebraic result.

LEMMA 2.2. *Let $u^\delta(x) = C_\delta(1 + |x|^2)^{-\frac{\delta}{2}}$ for $\delta, C_\delta > 0$. Then*

$$\mathcal{P}^+_{\lambda,\Lambda}(D^2u^\delta) \geq C_\delta \delta \lambda \frac{\beta - \delta - 2}{(1 + |x|^2)^{\frac{\delta}{2}+1}} .$$

Moreover, for $v^\nu(x) = \Theta|x|^{-\nu}$, $\Theta, \nu > 0$, *we have for* $x \in \mathbb{R}^N \setminus \{0\}$

$$\mathcal{P}^+_{\lambda,\Lambda}(D^2 v^\nu) = -\lambda\nu\Theta(\nu+1)|x|^{-\nu-2} + \nu\Theta\Lambda|x|^{-\nu-2}(N-1)\ .$$

PROOF. First, recall that if $f = f(|x|)$, the eigenvalues of $D^2 f$ are $f''(|x|)$, which is simple, $\frac{f'(|x|)}{|x|}$ with multiplicity $N-1$. Then, denoting by $f^\delta(|x|) = C_\delta(1+|x|^2)^{-\frac{\delta}{2}}$, we have

$$(f^\delta)'(|x|) = -C_\delta\delta(1+|x|^2)^{-\frac{\delta}{2}-1}|x|\ ,$$

$$\begin{aligned}(f^\delta)''(|x|) &= C_\delta\delta(\delta+2)|x|^2(1+|x|^2)^{-(\frac{\delta}{2}+2)} - C_\delta\delta(1+|x|^2)^{-(\frac{\delta}{2}+1)} \\ &= C_\delta\delta(1+|x|^2)^{-(\frac{\delta}{2}+2)}[(\delta+2)|x|^2 - (1+|x|^2)] \\ &= C_\delta\delta(1+|x|^2)^{-(\frac{\delta}{2}+2)}[(\delta+1)|x|^2 - 1]\ .\end{aligned}$$

In view of the definition of the Pucci's maximal operator

$$\mathcal{P}^+_{\lambda,\Lambda}(D^2 u) = -\lambda\sum_{e_k>0} e_k - \Lambda\sum_{e_k<0} e_k$$

we first address the case $|x|^2 \le \frac{1}{\delta+1}$, where $u^\delta(|x|)$ is concave and decreasing. Then, exploiting the fact that $\Lambda(N-1) = \lambda(\beta-1)$, we have

$$\mathcal{P}^+_{\lambda,\Lambda}(D^2u^\delta) = -\Lambda C_\delta\delta(1+|x|^2)^{-(\frac{\delta}{2}+2)}[(\delta+1)|x|^2-1] + \Lambda C_\delta\delta(1+|x|^2)^{-(\frac{\delta}{2}+1)}(N-1)$$

$$= C_\delta\delta\Lambda\left[\frac{N-1}{(1+|x|^2)^{\frac{\delta}{2}+1}} - \underbrace{\frac{(\delta+1)|x|^2-1}{(1+|x|^2)^{\frac{\delta}{2}+2}}}_{\ge 0}\right] \ge C_\delta\delta\frac{\Lambda(N-1)}{(1+|x|^2)^{\frac{\delta}{2}+1}} = C_\delta\delta\frac{\lambda(\beta-1)}{(1+|x|^2)^{\frac{\delta}{2}+1}}\ .$$

When instead $|x|^2 \ge \frac{1}{\delta+1}$ we write

$$\begin{aligned}\mathcal{P}^+_{\lambda,\Lambda}(D^2u^\delta) &= -\lambda C_\delta\delta(1+|x|^2)^{-(\frac{\delta}{2}+2)}[(\delta+1)|x|^2-1] + \Lambda C_\delta\delta(1+|x|^2)^{-(\frac{\delta}{2}+1)}(N-1) \\ &= C_\delta\delta\lambda\left[\frac{\beta-1}{(1+|x|^2)^{\frac{\delta}{2}+1}} - \frac{(\delta+1)|x|^2-1}{(1+|x|^2)^{\frac{\delta}{2}+2}}\right] \\ &= C_\delta\delta\lambda\left[\frac{(\beta-1)(1+|x|^2)-(\delta+1)|x|^2+1}{(1+|x|^2)^{\frac{\delta}{2}+2}}\right] \\ &= C_\delta\delta\lambda\left[\frac{(\beta-\delta-2)(1+|x|^2)+\beta}{(1+|x|^2)^{\frac{\delta}{2}+2}}\right] \\ &\ge C_\delta\delta\lambda\frac{\beta-\delta-2}{(1+|x|^2)^{\frac{\delta}{2}+1}}\ .\end{aligned}$$

where in the last inequality we used that $(\beta-\delta-2)|x|^2+\beta > (\beta-\delta-2)(1+|x|^2)$. In both cases, we get

$$\mathcal{P}^+_{\lambda,\Lambda}(D^2u^\delta) \ge C_\delta\delta\lambda\frac{\beta-\delta-2}{(1+|x|^2)^{\frac{\delta}{2}+1}}\ .$$

The second statement follows similarly, since $g^\nu(|x|) = \Theta|x|^{-\nu}$, $(g^\nu)' = -\Theta\nu|x|^{-\nu-1} < 0$ and $(g^\nu)'' = \nu(\nu+1)\Theta|x|^{-\nu-2} > 0$. □

We now state a few lemmata concerning the behavior of $m(R)$.

LEMMA 2.3. *The map $R \mapsto m(R)$ has the following properties.*

i) If $\mathcal{P}^+_{\lambda,\Lambda}(D^2u) \geq 0$ on $\mathbb{R}^N$, then $m(R)R^{\beta-2}$ is nondecreasing.

ii) If $\mathcal{P}^+_{\lambda,\Lambda}(D^2u) \geq |Du|^\gamma$ on $\mathbb{R}^N$, $m(R) \to 0$ as $R \to \infty$ and $\gamma \leq \frac{\beta}{\beta-1}$, then $m(R)R^\nu$ is nondecreasing for any $\nu \in (0, \beta-2)$ and R large enough.

PROOF. The proof of the two items follow similar lines. We start from *i)*, and follow [**32**, Theorem 4.1]. Fix $0 < R_1 < R$, we claim that the function

$$\Psi(x) = \frac{m(R_1) - m(R)}{R_1^{-(\beta-2)} - R^{-(\beta-2)}}(|x|^{-(\beta-2)} - R^{-(\beta-2)}) + m(R)$$

satisfies in the classical sense the equality $\mathcal{P}^+_{\lambda,\Lambda}(D^2\Psi) = 0$ for $|x| > R_1$. Indeed, setting $\Theta = \frac{m(R_1)-m(R)}{R_1^{-(\beta-2)}-R^{-(\beta-2)}}$, by Lemma 2.2 and the fact that $\Lambda(N-1) = \lambda(\beta-1)$ we obtain

$$\mathcal{P}^+_{\lambda,\Lambda}(D^2\Psi) = -\lambda(\beta-2)\Theta(\beta-1)|x|^{-\beta} + (\beta-2)\Theta\Lambda|x|^{-\beta}(N-1) = 0\ .$$

We then apply the comparison principle to the viscosity / classical inequalities

$$\mathcal{P}^+_{\lambda,\Lambda}(D^2u) \geq u^q > \mathcal{P}^+_{\lambda,\Lambda}(D^2\Psi)$$

in the annulus $A_{R_1,R} = \{R_1 < |x| < R\}$ (cf [**35**, Proposition 5]). Since $u \geq \Psi$ on $\partial A_{R_1,R}$ by construction we get $u \geq \Psi$ in $A_{R_1,R}$. Moreover, since $m(R) \geq 0$,

$$u(x) \geq \Psi(x) \geq \frac{m(R_1)\big(|x|^{-(\beta-2)} - R^{-(\beta-2)}\big)}{R_1^{-(\beta-2)} - R^{-(\beta-2)}},$$

and *i)* follows by letting $R \to \infty$.

The proof of *ii)* is a bit more delicate. As before, we need to show that

$$\Psi(x) = \frac{m(R_1) - m(R)}{R_1^{-\nu} - R^{-\nu}}(|x|^{-\nu} - R^{-\nu}) + m(R)$$

is a classical subsolution of $\mathcal{P}^+_{\lambda,\Lambda}(D^2\Psi) \leq |D\Psi|^\gamma$ in $A_{R_1,R}$. Setting $\Theta = \frac{m(R_1)-m(R)}{R_1^{-\nu}-R^{-\nu}}$ (which is positive, since $m(R)$ is decreasing), by the identity $|D\Psi|^\gamma = (\nu\Theta)^\gamma|x|^{-(\nu+1)\gamma}$ we have such an inequality provided that

$$|x|^{-\nu-2}\nu\Theta[-\lambda(\nu+1) + \lambda(\beta-1) - (\nu\Theta)^{\gamma-1}|x|^{-(\nu+1)\gamma+\nu+2}] \leq 0\ ,$$

namely

$$\Theta|x|^{-\nu+\frac{2-\gamma}{\gamma-1}} \geq \frac{\lambda^{\frac{1}{\gamma-1}}(\beta-\nu-2)^{\frac{1}{\gamma-1}}}{\nu}.$$

Under the condition $1 < \gamma \leq \frac{\beta}{\beta-1}$, it is immediate to verify that $\frac{2-\gamma}{\gamma-1} \geq \beta-2$, hence $-\nu + \frac{2-\gamma}{\gamma-1} > 0$. Recalling the definition of Θ, then the previous inequality holds true whenever $|x| \geq R_1$ and

$$m(R_1)R_1^{\frac{2-\gamma}{\gamma-1}} \geq \big(1 - (R_1/R)^\nu\big)\frac{\lambda^{\frac{1}{\gamma-1}}(\beta-\nu-2)^{\frac{1}{\gamma-1}}}{\nu} + m(R)R_1^{\frac{2-\gamma}{\gamma-1}}. \tag{2.2}$$

On one hand, by *i)*

$$m(R_1)R_1^{\frac{2-\gamma}{\gamma-1}} \geq m(R_1)R_1^{\beta-2} \geq m(1)$$

for all $R_1 \geq 1$. On the other hand, note first that we can assume $m(1) > 0$, otherwise by the strong maximum principle $u \equiv 0$ on $\mathbb{R}^N$ and the statement would be trivial. Since $m(R) \to 0$, $m(R)R_1^{\frac{2-\gamma}{\gamma-1}} \le \frac{m(1)}{2}$ for R large enough (depending on R_1 and $m(1)$) and thus

$$(1-(R_1/R)^\nu)\frac{\lambda^{\frac{1}{\gamma-1}}(\beta-\nu-2)^{\frac{1}{\gamma-1}}}{\nu} + m(R)R_1^{\frac{2-\gamma}{\gamma-1}} \le \frac{\lambda^{\frac{1}{\gamma-1}}(\beta-\nu-2)^{\frac{1}{\gamma-1}}}{\nu} + \frac{m(1)}{2} \le m(1)$$

provided that $\nu \in [\bar\nu, \beta-2)$, and $\bar\nu$ is sufficiently close to $\beta-2$ (depending on $\lambda, \gamma, m(1)$ only, and not on R_1). Therefore, (2.2) is verified, and thus Ψ is a subsolution as desired. Arguing by comparison as in *i)*, we obtain that $m(R)R^\nu$ is nondecreasing for $R \ge 1$ and such ν.

To extend the statement to any $\nu \in (0, \beta-2)$, note that

$$m(R_1)R_1^{\frac{2-\gamma}{\gamma-1}} \ge m(R_1)R_1^{\bar\nu}R_1^{\beta-2-\bar\nu} \ge m(1)R_1^{\beta-2-\bar\nu} \to \infty \quad \text{as } R_1 \to \infty,$$

and therefore (2.2) is verified whenever R_1 is large enough (now depending on $\lambda, \gamma, m(1), \beta, \nu$) and R is large enough (depending on R_1). We proceed as before by comparison and let $R \to \infty$ to get the statement. □

LEMMA 2.4. *Let u be a nonnegative solution to the viscosity inequality*

$$\mathcal{P}^+_{\lambda,\Lambda}(D^2u) \ge u^q$$

in $\mathbb{R}^N$. Then, the following bound from above holds

$$m(R) \le CR^{-\frac{2}{q-1}}, \; C > 0 \; . \tag{2.3}$$

PROOF. The result has been proved in [**32**, Theorem 4.1]. We sketch the proof here for the reader's convenience (in [**32**] some additional weights in the problem appear, but we do not have them here). For fixed $R > r$, we consider the test function

$$\varphi(x) = m(r)\left\{1 - \frac{[(|x|-r)^+]^3}{(R-r)^3}\right\} .$$

We observe that $\varphi \le 0 < u$ for $|x| \ge R$, while $\varphi(x) \equiv m(r) < u(x)$ for $|x| < r$. Since $\varphi(x) = u(x)$ at least at one point verifying $|x| = r$, we conclude that $u - \varphi$ achieves the nonpositive minimum in $\mathbb{R}^N$ at a certain point $\bar x$ in $\{x \in \mathbb{R}^N : r \le |x| < R\}$. Thus we can use φ as a test function in the definition of viscosity solution to find

$$\mathcal{P}^+_{\lambda,\Lambda}(D^2\varphi(\bar x)) \ge u^q(\bar x) .$$

By straightforward calculations as in Lemma 2.2 we get

$$\mathcal{P}^+_{\lambda,\Lambda}(D^2\varphi(\bar x)) = \frac{3\Lambda m(r)}{(R-r)^3}\left[2 + (N-1)\frac{(|x|-r)^+}{|x|}\right](|x|-r)^+$$

so that

$$\frac{3\Lambda m(r)}{(R-r)^3}\left[2 + (N-1)\frac{(|\bar x|-r)^+}{|\bar x|}\right](|\bar x|-r)^+ \ge u^q(\bar x) .$$

If $|\bar x| = r$, then we would find $u(\bar x) = 0$, which is a contradiction. So we may assume $r < |\bar x| < R$. In this case, we conclude

$$u^q(\bar x) \le \frac{3\Lambda(N+1)m(r)}{(R-r)^2} .$$

We then exploit the fact that $u(\bar{x}) \geq \min_{r<|x|<R} u(x) \geq m(R)$ and get

$$m^q(R) \leq \frac{3\Lambda(N+1)m(r)}{(R-r)^2} \quad \text{for } R > r\ .$$

In view of Lemma 2.3 we use that $m(r) \leq \frac{m(R)R^{\beta-2}}{r^{\beta-2}}$ and conclude

$$m^q(R) \leq \frac{3\Lambda(N+1)m(R)R^{\beta-2}}{r^{\beta-2}(R-r)^2} \quad \text{for } R > r\ .$$

We then take $r = R/2$ to get

$$m(R) \leq \frac{[3\Lambda(N+1)m(R)]^{\frac{1}{q}} 2^{\frac{1}{q}+\frac{\beta-2}{q}}}{R^{\frac{2}{q}}}\ ,$$

that yields (2.3). □

We are now ready to prove the main theorem of this section.

Proof of Theorem 2.1. Assume by contradiction that u is not constant, so $u > 0$ on $\mathbb{R}^N$ by the strong minimum principle, and $m(R_1) > 0$ for all R_1. Since u is a supersolution to (2.1), then it is also a supersolution to the equation $\mathcal{P}^+_{\lambda,\Lambda}(D^2u) \geq u^q$ on the whole $\mathbb{R}^N$, and also to $\mathcal{P}^+_{\lambda,\Lambda}(D^2u) \geq |Du|^\gamma$. Therefore, by Lemma 2.4 we have

$$m(R) \leq CR^{-\frac{2}{q-1}}. \tag{2.4}$$

Note that this implies $m(R) \to 0$ as $R \to \infty$. Therefore, we can apply Lemma 2.3, *ii)* to obtain for large $R > R_1$

$$m(R) \geq m(R_1)R_1^\nu R^{-\nu},$$

contradicting (2.4) when $\nu < \frac{2}{q-1}$ and $R \to \infty$. □

Some remarks and open questions are now in order.

Remark 2.5. The result in Theorem 2.1 is sharp. Indeed, when $\gamma > \frac{\beta}{\beta-1}$ and $q > \frac{\beta}{\beta-2}$, one can easily construct a counterexample to the Liouville property for the equation

$$\mathcal{P}^+_{\lambda,\Lambda}(D^2u) \geq u^q + |Du|^\gamma \text{ in } \mathbb{R}^N$$

of the form $v^\delta(x) = K_\delta(1+|x|^2)^{-\frac{\delta}{2}}$ for some suitable $K_\delta > 0$ provided that

$$\max\left\{\frac{2}{q-1}, \frac{2-\gamma}{\gamma-1}\right\} < \delta < \beta - 2\ . \tag{2.5}$$

Indeed, by Lemma 2.2 we have

$$\begin{aligned}\mathcal{P}^+_{\lambda,\Lambda}(D^2v^\delta) - (v^\delta)^q - |Dv^\delta|^\gamma &\geq (1+|x|^2)^{-\frac{\delta}{2}-1}\big[\lambda\delta K_\delta(\beta-\delta-2)- \\ &- (K_\delta)^q(1+|x|^2)^{-\frac{\delta}{2}q+\frac{\delta}{2}+1} - (\delta K_\delta)^\gamma|x|^\gamma(1+|x|^2)^{-(\frac{\delta}{2}+1)\gamma+\frac{\delta}{2}+1}\big]\ ,\end{aligned}$$

which is nonnegative whenever the inequalities in (2.5) are satisfied and K_δ is small enough. Indeed, when $-\frac{\delta}{2}q + \frac{\delta}{2} + 1 < 0$, i.e. $\delta > \frac{2}{q-1}$, we have

$$-(K_\delta)^q(1+|x|^2)^{-\frac{\delta}{2}q+\frac{\delta}{2}+1} \geq -K_\delta^q.$$

Moreover, using that $|x|^\gamma \le (1+|x|^2)^{\frac{\gamma}{2}}$, we have $-(\frac{\delta}{2}+\frac{1}{2})\gamma+\frac{\delta}{2}+1<0$ choosing $\delta > \frac{2-\gamma}{\gamma-1}$ and hence

$$-(\delta K_\delta)^\gamma |x|^\gamma (1+|x|^2)^{-(\frac{\delta}{2}+1)\gamma+\frac{\delta}{2}+1} \ge -(\delta K_\delta)^\gamma (1+|x|^2)^{-(\frac{\delta}{2}+\frac{1}{2})\gamma+\frac{\delta}{2}+1} \ge -(\delta K_\delta)^\gamma.$$

We are then left with the inequality

$$\begin{aligned}\mathcal{P}^+_{\lambda,\Lambda}(D^2 v^\delta) - (v^\delta)^q - |Dv^\delta|^\gamma &\ge (1+|x|^2)^{-\frac{\delta}{2}-1}[\lambda\delta K_\delta(\beta-\delta-2) - (K_\delta)^q - (\delta K_\delta)^\gamma] \\ &= K_\delta(1+|x|^2)^{-\frac{\delta}{2}-1}[\lambda\delta(\beta-\delta-2) - K_\delta^{q-1} - \delta^\gamma K_\delta^{\gamma-1}] \\ &\ge K_\delta(1+|x|^2)^{-\frac{\delta}{2}-1}[\lambda\delta(\beta-\delta-2) - 2\max\{K_\delta^{q-1}, \delta^\gamma K_\delta^{\gamma-1}\}].\end{aligned}$$

Therefore, if choose $\delta < \beta - 2$ we are free to take K_δ small enough so that the quantity $\lambda\delta(\beta-\delta-2) - 2\max\{K_\delta^{q-1}, \delta^\gamma K_\delta^{\gamma-1}\} \ge 0$.

REMARK 2.6. The previous remark shows that the theorem is sharp whenever $F = \mathcal{P}^+_{\lambda,\Lambda}$. In turn, for other operators, one could have the same Liouville property for a wider range of exponents q, γ. The most favorable case is indeed when $F = \mathcal{P}^-_{\lambda,\Lambda}$. With this choice, one could first deduce the Liouville property for classical solutions from linear results involving the Laplacian, that have been obtained in [**1**, **26**]: since

$$-\Lambda\Delta u \ge \mathcal{P}^-_{\lambda,\Lambda}(D^2 u) \ge u^q + |Du|^\gamma,$$

then u has to be constant whenever

$$q > \frac{N}{N-2} \quad \text{and} \quad \gamma \le \frac{N}{N-1}, \quad \left(\text{or } q \le \frac{N}{N-2} \text{ and any } \gamma > 0\right),$$

where $N > 2$. However, this range can be further improved. Indeed, if $\mathcal{P}^-_{\lambda,\Lambda}(D^2 u) \ge 0$, arguing as in Lemma 2.3 one can prove that $R \mapsto m(R)R^{\alpha-2}$ is nondecreasing, where $\alpha = \frac{\lambda}{\Lambda}(N-1)+1$. Then, one can retrace the same proof of Theorem 2.1 replacing β with α, and obtain the result when

$$q > \frac{\alpha}{\alpha-2} \quad \text{and} \quad \gamma \le \frac{\alpha}{\alpha-1}, \quad \left(\text{or } q \le \frac{\alpha}{\alpha-2} \text{ and any } \gamma > 0\right),$$

which is even a larger region for (q,γ) since $\frac{\alpha}{\alpha-1} > \frac{N}{N-1} > \frac{\beta}{\beta-1}$ (at least when $\Lambda > \lambda$).

REMARK 2.7. One can generalize Theorem 2.1 to fully nonlinear equations of the form

$$F(x, D^2 u) \ge H(x, u, Du) \text{ in } \mathbb{R}^N,$$

with

$$H(x,u,Du) \ge a(x)u^q + b(x)|Du|^\gamma$$

where e.g. $a(x) = |x|^{-\sigma}$, $\sigma > -2$, and $b(x) \ge b_0$. In this case the range of exponents in Theorem 2.1 becomes $\gamma \le \frac{\beta}{\beta-1}$ and $q > \frac{\beta+\sigma}{\beta-2}$ (the case $q \le \frac{\beta+\sigma}{\beta-2}$ being a consequence of [**32**, Theorem 4.1]). One can even replace u^q with $f = f(u)$, where $f : (0,\infty) \to \mathbb{R}$ is a continuous function verifying

$$\liminf_{s\to 0} \frac{f(s)}{s^q} > 0.$$

Finally, we believe that the result in Theorem 2.1 can be generalized to exterior domains of $\mathbb{R}^N$, i.e. when the equation is posed on $\mathbb{R}^N \backslash B$ where $B \subset \mathbb{R}^N$ is any ball, $N > 2$, following a similar scheme of proof. Nevertheless, some arguments become more delicate; for example, it is no longer true in general that $\min_{R_0 \le |x| \le R} u(x) =$

$\min_{|x|=R} u(x)$ on exterior domains. This will be the matter of a future research, together with the treatment of a general case, as summarized in the next open problem.

OPEN PROBLEM 2.8. Do the results in Theorem 2.1 extend to general Bellman-Isaacs equations as in [**6**], that is, of the form $\sup_\sigma \inf_\eta \mathcal{L}^{\sigma\eta}u = 0$, where $\mathcal{L}^{\sigma\eta}$ is a family of linear uniformly elliptic operators, and to exterior domains? What happens when the gradient term in (2.1) has the opposite sign as in [**1**], i.e. $H(u, Du) = u^q - |Du|^\gamma$?

Similar Liouville properties to those outlined in this section have been recently addressed in [**11**] for problems driven by fractional Laplacians. We now formulate the following

OPEN PROBLEM 2.9. Do the results of this section extend to problems driven by fully nonlinear nonlocal operators as in [**34**], namely for these problems involving

$$\mathcal{P}^+u = \sup_{K\in\mathcal{F}} -\int_{\mathbb{R}^N} (u(x+y) + u(x-y) - 2u(x))K(y)\,dy\ ,$$

$$\mathcal{P}^-u = \inf_{K\in\mathcal{F}} -\int_{\mathbb{R}^N} (u(x+y) + u(x-y) - 2u(x))K(y)\,dy\ ,$$

where $\mathcal{F} = \left\{K : \frac{\lambda}{|y|^{N+2s}} \le K(y) \le \frac{\Lambda}{|y|^{N+2s}}\ , K(y) = K(-y)\right\}$, $\Lambda \ge \lambda > 0$, $s \in (0,1)$?

3. Failure of the Liouville property for nonlinearities involving the product of powers of the unknown and its gradient

We address in this section the failure of the Liouville property for inequalities involving *products* of zero-th and first order nonlinearities, that is,

$$F(x, D^2u) \ge u^q|Du|^\gamma \quad \text{in } \mathbb{R}^N,\ q \ge 0\ , \gamma > 1\ .$$

with F continuous, uniformly elliptic and satisfying $F(x, 0) = 0$. Nonlinearities of the form $u^q|Du|^\gamma$ appear, for example, after a change of variable in the generalized porous-medium equations (see e.g. the introduction in [**27**]). By the uniform ellipticity, any solution to the above inequality satisfies in the viscosity sense

$$\mathcal{P}^+_{\lambda,\Lambda}(D^2u) \ge u^q|Du|^\gamma \quad \text{in } \mathbb{R}^N,\ q \ge 0\ , \gamma > 1\ . \tag{3.1}$$

The validity of the Liouville property in the linear case $F = -\Delta$ has been shown in [**23, 26**], and requires $(N-1)\gamma + (N-2)q \le N$. For the nonlinear counterpart, we state the following necessary condition.

THEOREM 3.1. *Let $N > 2$. Then, there exist non-constant nonnegative classical solutions to the inequality* (3.1) *when $(\beta-1)\gamma+(\beta-2)q > \beta$, where $\beta = \frac{\Lambda}{\lambda}(N-1)+1$. In particular, when $q = 0$ there exist non-constant entire solutions to $\mathcal{P}^+_{\lambda,\Lambda}(D^2u) \ge |Du|^\gamma$ in $\mathbb{R}^N$ when $\gamma > \frac{\beta}{\beta-1}$.*

PROOF OF THEOREM 3.1. Consider $v(x) = C_\delta(1 + |x|^2)^{-\frac{\delta}{2}}$ for $\delta, C_\delta > 0$ to be determined. By Lemma 2.2 we have

$$\mathcal{P}^+_{\lambda,\Lambda}(D^2v) - v^q|Dv|^\gamma \ge C_\delta\delta\lambda\frac{\beta-\delta-2}{(1+|x|^2)^{\frac{\delta}{2}+1}} - \frac{(C_\delta\delta)^\gamma}{(1+|x|^2)^{(\frac{\delta}{2}+1)\gamma-\frac{\gamma}{2}}}\frac{C_\delta^q}{(1+|x|^2)^{\frac{\delta}{2}q}}$$

The right-hand side of the above inequality is bigger than or equal to zero when

$$\begin{cases} \left(\frac{\delta}{2}+1\right) \leq \left(\frac{\delta}{2}+1\right)\gamma - \frac{\gamma}{2} + \frac{\delta}{2}q \\ \beta > \delta+2 \\ \lambda\delta C_\delta(\beta-\delta-2) - (C_\delta\delta)^\gamma C_\delta^q \geq 0 \ . \end{cases}$$

In particular, coupling the first two constraints it is easy to see that one gets $(\beta-2)q+(\beta-1)\gamma > \beta$, while the fact that $\beta-\delta-2>0$ allows to choose a suitable (small) positive constant $C_\delta > 0$. For example, when $q=0$ the equality in the first constraint is attained e.g. for $\delta = \frac{2-\gamma}{\gamma-1}$, while the second inequality gives $\gamma > \frac{\beta}{\beta-1}$, so that the function $v(x) = C_\delta(1+|x|^2)^{-\frac{2-\gamma}{2(\gamma-1)}}$ works as a counterexample for the Liouville property for C_δ small enough (so that the third inequality is satisfied). □

REMARK 3.2. The same kind of function serves as a counterexample for equations posed on exterior domains of $\mathbb{R}^N$. It is straightforward to check that the (singular) function $v(x) = c|x|^{-\frac{2-\gamma}{\gamma-1}}$ in $\mathbb{R}^N\setminus\{0\}$ is a classical solution to $\mathcal{P}^+_{\lambda,\Lambda}(D^2v) \geq |Dv|^\gamma$ in $\mathbb{R}^N\setminus\{0\}$ for $\gamma > \frac{\beta}{\beta-1}$.

We now formulate the following conjecture, that extends [**26**, Theorem 7.1], [**23**, Corollary 3] and [**18**, Theorem 2.1] to the fully nonlinear setting.

CONJECTURE 3.3. Any nonnegative solution to $\mathcal{P}^+_{\lambda,\Lambda}(D^2u) \geq u^q|Du|^\gamma$ in $\mathbb{R}^N$ is constant provided that $(\beta-2)q+(\beta-1)\gamma \leq \beta$, $\gamma > 1$, $q \geq 0$. This would agree with the linear case, since $\beta \to N$ when $\lambda/\Lambda \to 1$.

Note that by a suitable change of variables, the problem can be reduced to the case $q=0$. The heuristic calculation goes as follows (we proceed as in [**18**, Theorem 2.1]): under the assumption

$$(\beta-2)q+(\beta-1)\gamma < \beta \tag{3.2}$$

we let $u = v^b$, for some $b \in \mathbb{R}$ satisfying $b(b-1) > 0$ to be later determined. Moreover, to simplify the calculation, suppose that u, v are smooth functions. We have

$$Du = bv^{b-1}Dv \ , |Du|^2 = b^2v^{2(b-1)}|Dv|^2 \ , |Du|^\gamma = b^\gamma v^{\gamma(b-1)}|Dv|^\gamma$$

and

$$D^2u = b(b-1)v^{b-2}Dv\otimes Dv + bv^{b-1}D^2v \ .$$

We then conclude by the inequality $\mathcal{P}^+_{\lambda,\Lambda}(M+N) \leq \mathcal{P}^+_{\lambda,\Lambda}(M) + \mathcal{P}^+_{\lambda,\Lambda}(N)$ that

$$\begin{aligned} \mathcal{P}^+_{\lambda,\Lambda}(D^2u) &= \mathcal{P}^+_{\lambda,\Lambda}(b(b-1)v^{b-2}Dv\otimes Dv + bv^{b-1}D^2v) \\ &\leq \mathcal{P}^+_{\lambda,\Lambda}(b(b-1)v^{b-2}Dv\otimes Dv) + \mathcal{P}^+_{\lambda,\Lambda}(bv^{b-1}D^2v) \\ &= -b(b-1)v^{b-2}\lambda|Dv|^2 + bv^{b-1}\mathcal{P}^+_{\lambda,\Lambda}(D^2v) \ . \end{aligned}$$

From the inequality $\mathcal{P}^+_{\lambda,\Lambda}(D^2u) \geq u^q|Du|^\gamma$ we then get

$$b\mathcal{P}^+_{\lambda,\Lambda}(D^2v) \geq \lambda b(b-1)\frac{|Dv|^2}{v} + b^\gamma v^s|Dv|^\gamma$$

for $s = 1-\gamma+b(q+\gamma-1)$. We now use the generalized Young's inequality (s will be positive, see below) to find that

$$|Dv|^{\frac{2s+\gamma}{s+1}} = \left(\frac{|Dv|^2}{v}\right)^{\frac{s}{s+1}} v^{\frac{s}{s+1}}|Dv|^{\frac{\gamma}{s+1}} \leq \varepsilon^{\frac{s+1}{s}}\frac{|Dv|^2}{v} + \frac{1}{\varepsilon^{s+1}}|Dv|^\gamma v^s$$

for any $\varepsilon > 0$. We choose $\varepsilon^{-1-s} = b^q$ to find that for $c > 0$ satisfying $c^{\frac{s+1}{s}} = \lambda b^{1-\frac{\gamma}{s}}(b-1)$ it holds

$$\mathcal{P}^+_{\lambda,\Lambda}(D^2v) \geq c|Dv|^z \text{ in } \mathbb{R}^N , \tag{3.3}$$

where

$$z = \frac{2s+\gamma}{s+1} = \frac{2-\gamma+2b(q+\gamma-1)}{2-\gamma+b(q+\gamma-1)} .$$

For $\gamma+q-1>0$, we take $b = 1+\delta$, $\delta > 0$, so that $b(b-1) > 0$. Then, we observe that $s = \gamma+\delta(\gamma+q-1) > 0$ and $s = 1-\gamma+\delta(q+\gamma-1) > 1-\gamma$, which implies $z > 1$. By the assumption (3.2) we can take δ small enough to find $(\beta-2)s+(\beta-1)\gamma < \beta$. Indeed

$$\begin{aligned}(\beta-2)s+(\beta-1)\gamma &< (\beta-2)s+\beta-(\beta-2)q \\ &= (\beta-2)(s-q)+\beta = (\beta-2)(q+\gamma-1)(\delta-1)+\beta ,\end{aligned}$$

which is equivalent to

$$z < \frac{\beta}{\beta-1} , \tag{3.4}$$

since

$$\begin{aligned}(\beta-2)s+(\beta-1)\gamma < \beta &\implies 2s(\beta-1)+(\beta-1)\gamma < \beta(s+1) \\ &\implies (2s+\gamma)(\beta-1) < \beta(s+1) .\end{aligned}$$

Therefore, the Liouville property for (3.1) reduces to the one for (3.3), which we conjecture to be true whenever (3.4) holds.

A similar transformation can be carried out for the equation $\mathcal{P}^-_{\lambda,\Lambda}(D^2u) \geq u^q|Du|^\gamma$, by exploiting the inequality $\mathcal{P}^-_{\lambda,\Lambda}(M+N) \leq \mathcal{P}^-_{\lambda,\Lambda}(M) + \mathcal{P}^+_{\lambda,\Lambda}(N)$. Then, condition (3.2) becomes $(\alpha-2)q+(\alpha-1)\gamma < \alpha$, yielding $z < \frac{\alpha}{\alpha-1}$.

REMARK 3.4. Following Remark 2.6, when $F = \mathcal{P}^-$, one can obtain the Liouville property for nonnegative solutions to the viscosity inequality

$$\mathcal{P}^-_{\lambda,\Lambda}(D^2u) \geq u^q|Du|^\gamma \text{ in } \mathbb{R}^N$$

from the known results for the Laplacian [**26**, Theorem 7.1], [**18**, Theorem 2.1], [**23**, Corollary 3] via the inequality $-\Lambda\Delta u \geq \mathcal{P}^-_{\lambda,\Lambda}(D^2u)$ when $(N-2)q+(N-1)\gamma \leq N$. However, in this case one should expect the Liouville property when $(\alpha-2)q+(\alpha-1)\gamma \leq \alpha$, $\alpha = \frac{\lambda}{\Lambda}(N-1)+1$, which yields a wider region than $(N-2)q+(N-1)\gamma \leq N$.

REMARK 3.5. It is worth remarking that related Liouville properties to those in this section were established in [**20**, Theorem 3.1], where it was proved that if $u \in C(\mathbb{R}^N)$ is a nonnegative viscosity solution (in an appropriate sense) of

$$\mathcal{P}^+_{\lambda,\Lambda}(D^2u)|Du|^\eta \geq h(x)u^q ,$$

where

$$\eta > -1 \text{ , } h(x) = h|x|^\sigma \text{ for } |x| \text{ large}, h > 0 \text{ and } \sigma > -(\eta+2) ,$$

and

$$0 < q \leq \frac{1+\sigma+(\eta+1)(\beta-1)}{\beta-2} ,$$

then $u \equiv 0$. This is accomplished by modifying the proof in Lemma 2.4 to achieve the bound $m(R)R^{\beta-2} \leq CR^{\beta-2-\frac{\eta+2+\sigma}{q-(1+\eta)}}$ through a different choice of the test function φ, which leads to the desired contradiction since $\beta-2-\frac{\eta+2+\sigma}{q-(1+\eta)} < 0$. Therefore, if one takes $\sigma = 0$, $\gamma := -\eta < 1$, it is possible to deduce the Liouville property for suitable viscosity solutions of the inequality

$$\mathcal{P}^{+}_{\lambda,\Lambda}(D^2u) \geq u^q|Du|^\gamma \text{ in } \mathbb{R}^N$$

in the following range

$$(\beta-2)q + (\beta-1)\gamma \leq \beta \ , \ 0 < \gamma < 1 \ , \ q > 0 \ , \ N > 2 \ .$$

that however is outside the framework of the previous Conjecture 3.3. In particular, one cannot deduce the result in the case $q = 0$ (namely in the case of the absence of zero-th order terms), that remains at this stage an open problem even in the sublinear regime of the gradient nonlinearity for equations driven by fully nonlinear operators. Finally, observe that a suitable modification of the scheme of proof in [**20**, Theorem 3.1] leads to the improved Liouville property for the inequality

$$\mathcal{P}^{-}_{\lambda,\Lambda}(D^2u) \geq u^q|Du|^\gamma \text{ in } \mathbb{R}^N \ , u \geq 0 \ , .$$

under the assumptions

$$(\alpha-2)q + (\alpha-1)\gamma \leq \alpha \ , \ 0 < \gamma < 1 \ , \ q > 0 \ , \ N > 2 \ , \ \alpha = \frac{\lambda}{\Lambda}(N-1)+1 \ .$$

REMARK 3.6. A partial result for $F = \mathcal{P}^{-}_{\lambda,\Lambda}$, without any restriction on q, γ could be deduced exploiting that nonnegative solutions to $\mathcal{P}^{-}_{\lambda,\Lambda}(D^2u) \geq u^q|Du|^\gamma$ are also solutions to $\mathcal{P}^{-}_{\lambda,\Lambda}(D^2u) \geq 0$ under the assumption $N \leq \frac{\Lambda}{\lambda}+1$ through [**32**, Theorem 3.2]. For general uniformly elliptic operators, the Liouville property holds for any $q, \gamma \geq 0$ in the one dimensional case, see Remark 5.2 below.

The previous result could be also the starting point to address the following generalization of [**37, 75**] to systems of fully nonlinear inequalities with first-order terms having super-linear growth.

OPEN PROBLEM 3.7. Does the Liouville property hold for the system

$$\begin{cases} F(x, D^2u) \geq v^{q_1}|Dv|^{\gamma_1} & \text{in } \mathbb{R}^N \ , \\ F(x, D^2v) \geq u^{q_2}|Du|^{\gamma_2} & \text{in } \mathbb{R}^N \ , \end{cases}$$

where F is uniformly elliptic and the exponents $q_1, q_2, \gamma_1, \gamma_2$ are as in the previous sections?

We now mention an open problem for degenerate operators:

OPEN PROBLEM 3.8. Let $\mathcal{X}$ be a family of Hörmander vector fields, $D_{\mathcal{X}}$ be the horizontal gradient in $\mathbb{R}^m$ and $(D^2_{\mathcal{X}})^*$ the symmetrized horizontal Hessian in Sym_m, $m \leq N$, (see Section 6 for precise definitions). Do the aforementioned Liouville properties extend to degenerate problems such as

$$\mathcal{P}^{\pm}_{\lambda,\Lambda}((D^2_{\mathcal{X}}u)^*) \pm u^q|D_{\mathcal{X}}u|^\gamma \geq u^p \text{ in } \mathbb{R}^N$$

for some values of q, γ, p?

4. Fully nonlinear equations with superlinear gradient growth perturbed by drift terms

We now consider Liouville properties for viscosity solutions to fully nonlinear equations of the form

$$F(x, D^2u) - b(x) \cdot Du = H(Du) \text{ in } \mathbb{R}^N, \tag{4.1}$$

where F satisfies the same assumptions of the previous sections and b is a confining vector field. Our results will be concerned with Hamiltonians $H(Du)$ with power-like growth. We will show that, under suitable assumptions on b, the drift term is strong enough to provide the Liouville property even when H has negative sign.

We assume that b is locally Lipschitz in x, namely for all $R > 0$ there exists $K_R > 0$ such that

$$\sup_{|x|,|y| \leq R} |b(x) - b(y)| \leq K_R |x - y| \ . \tag{4.2}$$

As mentioned in [**7**], a simple example of vector field satisfying the assumptions in this section are perturbation of Ornstein-Uhlenbeck velocity fields with a lower-order perturbation such as

$$b(x) = \nu(m - x) + h(x) \ , \nu > 0 \ , m \in \mathbb{R}^N \ , \lim_{|x|\to\infty} \frac{h(x) \cdot x}{|x|^2} = 0.$$

The main result of the section is the following

THEOREM 4.1. *Let F be continuous, uniformly elliptic and such that $F(x, 0) = 0$, $H(Du) = A|Du|^\gamma$, and either*

$$A > 0, \quad \textit{and} \quad \gamma > 0$$

or

$$A \in \mathbb{R}, \quad \textit{and} \quad \gamma = 2.$$

Assume also that b satisfies (4.2) *and*

$$\limsup_{|x|\to\infty} b(x) \cdot x < \lambda - \Lambda(N - 1) \ . \tag{4.3}$$

Let u be a supersolution to (4.1) *bounded from below. Then u is constant.*

The result shows that the presence of a suitable drift term forces the Liouville property, independently of the positive power γ in the equation. We recall once more that when $H(Du) = A|Du|^\gamma$ and $b \equiv 0$, one needs $\gamma \leq \frac{N}{N-1} (< 2)$, as shown in [**26**, Theorem 7.1]. Nevertheless, one can drop this assumption by requiring the velocity field b to point toward the origin for large $|x|$, and having a sufficiently large inward component. Such a condition is by no means new (it can be found in e.g. [**7**] or [**9**] and references therein). The main contribution of Theorem 4.1 is that (4.3) is strong enough to yield the Liouville property even when the super-linear gradient term has a *repulsive* behavior, for example when $H(Du) = -|Du|^2$, where the sign is negative.

The proof of our result will be based on the following crucial lemma, concerning the Liouville property for (4.1) when $H \equiv 0$. We will rely on the existence of a sort of "Lyapunov function", which in our nonlinear setting reads as the existence of an explosive subsolution to $\mathcal{P}^+_{\lambda,\Lambda}(D^2u) - b(x) \cdot Du = 0$, see (4.4) below. While the argument might look different with respect to the one that has been used in previous sections, we are still controlling from below the behavior of u with suitable subsolutions, which are somehow playing the role of the fundamental solutions in Section 2. Its proof builds upon strategies already appeared in e.g. [**9**, Theorem 2.1] for classical solutions to linear equations and [**7**, Theorem 2.1], [**9**, Proposition 3.1]. Note that the one-side bound can be relaxed, cf Remark 4.4 below.

LEMMA 4.2. *Under the assumptions on b of Theorem 4.1, let u be a supersolution to*

$$E(x, Du, D^2u) := \mathcal{P}^+_{\lambda,\Lambda}(D^2u) - b(x) \cdot Du \geq 0 \quad \text{in } \mathbb{R}^N$$

bounded from below. Then u is constant.

The proof is based on the following idea: first, one proves that there exists a compact set $K \subset \mathbb{R}^N$ such that $\inf_{\mathbb{R}^N} u = \min_K u$. Then, the function $z = u - \min_K u$ satisfies the viscosity inequality $E(x, Du, D^2u) \geq 0$ in $\mathbb{R}^N$, while z vanishes somewhere in K. Therefore, $z \equiv 0$ by the strong minimum principle, and hence u is constant. We prove this result through several steps, with the aid of a Lyapunov function, as outlined in the next proof. We premise the following well-known property, known as *transitivity of the viscosity inequalities.*

LEMMA 4.3. *Let u be satisfying the inequality $\mathcal{P}^+_{\lambda,\Lambda}(D^2u) \geq f$ in the viscosity sense and v be such that the inequality $\mathcal{P}^+_{\lambda,\Lambda}(D^2v) \leq g$ holds in the viscosity sense. Then, $w = u - \xi v$, $\xi > 0$, satisfies $\mathcal{P}^+_{\lambda,\Lambda}(D^2w) \geq f - \xi g$ in the viscosity sense.*

PROOF. The proof is simple when one of the involved functions is smooth, cf e.g. [**24**, Lemma 2.12]. A more general statement can be found in [**24**, Theorem 5.3] and [**6**, p. 745], while a proof for Hamilton-Jacobi-Bellman operators can be found in [**7**, Theorem 2.1-Step 3] or [**9**, Proposition 3.1]. □

PROOF.

STEP 1 (Existence of a Lyapunov function). We first observe that there exists $\bar{R} \geq 0$ such that the function $w(x) = -\log|x|$ satisfies $w \in C^2(\mathbb{R}^d \backslash \{0\})$,

$$\lim_{|x|\to\infty} w(x) = -\infty \quad \text{and} \quad E(x, Du, D^2u) \leq 0 \text{ in } \mathbb{R}^N \backslash \overline{B}(0, \bar{R}). \tag{4.4}$$

Indeed,

$$\mathcal{P}^+_{\lambda,\Lambda}(D^2w) - b(x) \cdot Dw = \frac{\Lambda(N-1)}{|x|^2} - \frac{\lambda}{|x|^2} + b(x) \cdot \frac{x}{|x|^2}$$

and (4.3) implies the (last) claim outside a large ball of radius $\bar{R}$.

STEP 2. For $\xi > 0$, we set $v_\xi := u - \xi w$. Then, v_ξ is continuous on the set $\{x \in \mathbb{R}^N : |x| \geq \bar{R}\}$. Moreover, by Lemma 4.3, we have that the following inequality holds in the viscosity sense

$$E(x, Dv_\xi, D^2v_\xi) \geq 0 \text{ for every } x \text{ such that } |x| > \bar{R}.$$

Define $c_\xi = \min_{|x|=\bar{R}} v_\xi$. Since u is bounded below, we have

$$\lim_{|x|\to\infty} v_\xi(x) = +\infty ,$$

and hence there exists $K_\xi > \bar{R}$ such that $v_\xi > c_\xi$ for every x such that $|x| \geq K_\xi$.

STEP 3. We use the comparison principle on $A_{R,K_\xi} := \{x \in \mathbb{R}^N : \bar{R} < |x| < K_\xi\}$, cf [**7**, **8**] and the references therein, to conclude

$$\min_{A_{R,K_\xi}} v_\xi(x) = \min_{\{x\in\mathbb{R}^N : |x|=\bar{R} \text{ or } |x|=K_\xi\}} v_\xi(x) = c_\xi .$$

Since $v_\xi > c_\xi$ for $|x| \geq K_\xi$, we get for all $|y| \geq \bar{R}$,

$$v_\xi(y) \geq c_\xi = \min_{\{x\in\mathbb{R}^N:|x|=\bar{R}\}} u + \xi \max_{\{x\in\mathbb{R}^N:|x|=\bar{R}\}} w \, .$$

We let $\xi \to 0$ and conclude

$$u(y) \geq \min_{\{x\in\mathbb{R}^N:|x|=\bar{R}\}} u \, , |y| \geq \bar{R} \, .$$

On the other hand, the comparison principle applied on the ball $B(0,\bar{R})$ leads to

$$u(y) \geq \min_{\{x\in\mathbb{R}^N:|x|=\bar{R}\}} u \, , |y| < \bar{R} \, .$$

Combining the above assertions, we conclude that u attains its minimum at some point of $\partial B(0,\bar{R})$, so that u is constant by the strong minimum principle [**8**, **44**]. □

We can now prove the main theorem of this section.

PROOF OF THEOREM 4.1. Let us start with the case $A > 0, \gamma > 0$. In such case, it is sufficient to observe that u is a supersolution bounded from below to

$$\mathcal{P}^+_{\lambda,\Lambda}(D^2u) - b(x)\cdot Du \geq A|Du|^\gamma \quad \text{in } \mathbb{R}^N$$

and in turn to

$$E(x, Du, D^2u) \geq 0 \text{ in } \mathbb{R}^N \, ,$$

so it is sufficient to apply Lemma 4.2 to obtain the desired result.

Let us now consider the case $H(Du) = A|Du|^2$, and assume $A = -1$ for simplicity. Then u solves in the viscosity sense

$$\mathcal{P}^+_{\lambda,\Lambda}(D^2u) + |Du|^2 - b(x)\cdot Du \geq 0 \text{ in } \mathbb{R}^N \, .$$

By the nonlinear Hopf-Cole transform, the function

$$v(x) = \lambda(1 - e^{-\frac{1}{\lambda}u(x)})$$

is a bounded below viscosity solution to the inequality

$$E(x, Dv, D^2v) = \mathcal{P}^+_{\lambda,\Lambda}(D^2v) - b(x)\cdot Dv \geq 0 \text{ in } \mathbb{R}^N \, .$$

To see this, assume first that u is smooth. We have

$$Dv = Du\, e^{-\frac{1}{\lambda}u}, \qquad D^2v = e^{-\frac{1}{\lambda}u}D^2u - \frac{1}{\lambda}Du\otimes Du\, e^{-\frac{1}{\lambda}u} \, .$$

Then, exploiting the inequality $\mathcal{P}^+_{\lambda,\Lambda}(M+N) \geq \mathcal{P}^+_{\lambda,\Lambda}(M) + \mathcal{P}^-_{\lambda,\Lambda}(N)$, we get

$$\begin{aligned}
\mathcal{P}^+_{\lambda,\Lambda}(D^2v) - b(x)\cdot Dv &= \mathcal{P}^+_{\lambda,\Lambda}\left(e^{-\frac{1}{\lambda}u}D^2u - \frac{1}{\lambda}Du\otimes Du\, e^{-\frac{1}{\lambda}u}\right) - b(x)\cdot Du\, e^{-\frac{1}{\lambda}u} \\
&\geq \mathcal{P}^+_{\lambda,\Lambda}(D^2u)e^{-\frac{1}{\lambda}u} + \mathcal{P}^-_{\lambda,\Lambda}\left(-\frac{1}{\lambda}Du\otimes Du\, e^{-\frac{1}{\lambda}u}\right) - b(x)\cdot Du\, e^{-\frac{1}{\lambda}u} \\
&= \mathcal{P}^+_{\lambda,\Lambda}(D^2u)e^{-\frac{1}{\lambda}u} - \frac{1}{\lambda}\mathcal{P}^+_{\lambda,\Lambda}\left(Du\otimes Du\right)e^{-\frac{1}{\lambda}u} - b(x)\cdot Du\, e^{-\frac{1}{\lambda}u} \\
&= (\mathcal{P}^+_{\lambda,\Lambda}(D^2u) + |Du|^2 - b(x)\cdot Du)e^{-\frac{1}{\lambda}u} \, ,
\end{aligned}$$

namely

$$\frac{\mathcal{P}^+_{\lambda,\Lambda}(D^2v) - b(x)\cdot Dv}{1 - v/\lambda} \geq \mathcal{P}^+_{\lambda,\Lambda}(D^2u) + |Du|^2 - b(x)\cdot Du \, .$$

When u is just a viscosity supersolution, suppose by contradiction that v does not satisfy $\mathcal{P}^+_{\lambda,\Lambda}(D^2v)-b(x)\cdot Dv\geq 0$ in the viscosity sense. Then, there exists a smooth function ψ such that $v-\psi$ attains a local minimum at some $x_0\in A$, and

$$\mathcal{P}^+_{\lambda,\Lambda}(D^2\psi)-b(x)\cdot D\psi<0 \text{ in } A\ .$$

We then define $\varphi(x)=-\lambda\log\left(1-\frac{\psi(x)}{\lambda}\right)$ in the above inequality to get a contradiction with the fact that u solves in the viscosity sense the initial equation, and $u-\varphi$ attains a local minimum at x_0. Hence, we can again apply Lemma 4.2. □

REMARK 4.4. One can relax the one-side bound from below on u by the weaker growth condition $\liminf_{|x|\to\infty}\frac{u(x)}{\log|x|}\geq 0$. Clearly, the latter condition is satisfied if u is bounded below.

Furthermore, one can also get a sufficient condition for the Liouville property in the presence of zero-th order terms such as $c(x)u$, $c>0$, by further imposing a two-side bound on the solution, see e.g. [**9**] or [**71**], where the property is called λ-Liouville property. In the case $\lambda=\Lambda=1$, this implies that $\mathbb{R}^N$ is stochastically complete. The drift term can be also replaced with a more general concave Hamiltonian such as

$$H_{\text{conc}}(x,u,Du)=\inf_{\alpha\in A}\{c^\alpha(x)u-b^\alpha(x)\cdot Du\}\ ,$$

where A is a metric space, under appropriate assumptions on b^α, c^α, cf [**7**, **9**].

Note that the theorem can be extended in an obvious manner to non-negative Hamiltonians. On the other hand, the power-like case $H=A|Du|^\gamma$, with *negative* A and $\gamma\neq 2$ is much more delicate. One may conjecture that the Liouville property holds as in the quadratic case $\gamma=2$, but we do not know how to prove it at this stage.

We conclude by saying that some Liouville properties for different equations posed on Riemannian manifolds through similar methods have been obtained for solutions to linear PDEs satisfying two-side bounds in e.g. [**45**, Section 13] and [**59**], see also [**72**] for fully nonlinear problems, where the existence of a Lyapunov function is required. Instead, other results for solutions bounded from one-side to nonlinear partial differential inequalities driven by the p-Laplacian have been established in [**71**, Section 4] and recently in [**60**, Theorem 1.2].

REMARK 4.5. Other sufficient conditions implying the Liouville property can be obtained by considering different negative power-like functions, as $w(x)=-\frac{|x|^2}{2}$.

REMARK 4.6. We show that the condition (4.3) imposed on the velocity field is optimal for the Liouville property, at least for the class of operators $\mathcal{P}^++b(x)\cdot D$, with the velocity field b satisfying the above assumptions. In the nonlinear setting, the claim that the validity of the Liouville property implies the existence of Lyapunov functions has been matter of recent research (see e.g. [**59**] and references therein). Here, we obtain such implication (and therefore the equivalence) for a family of problems involving $\mathcal{P}^+$ and Ornstein-Uhlenbeck drifts.

Note first that condition (4.3) can be equivalently rewritten as

$$\limsup_{|x|\to\infty}\{b(x)\cdot x\}<\lambda(2-\beta)\ .$$

Then, for $\delta>0$ we take $u(x)=(1+|x|^2)^{-\frac{\delta}{2}}=:f(|x|)$. Computations in Lemma 2.2 with $C_\delta=1$ give (using that $\Lambda(N-1)=\lambda(\beta-1)$ and the fact that for

$|x|^2 < 1/(\delta+1)$ one has $\Lambda[1-(\delta+1)|x|^2] > \lambda[1-(\delta+1)|x|^2])$

$$\mathcal{P}^+_{\lambda,\Lambda}(D^2u) \geq \delta\lambda \left[\frac{\beta-1}{(1+|x|^2)^{\frac{\delta}{2}+1}} - \frac{(\delta+1)|x|^2-1}{(1+|x|^2)^{\frac{\delta}{2}+2}} \right] .$$

Hence, for $b(x) = \lambda(2-\beta+\delta)\frac{x}{1+|x|^2}$, we have

$$\mathcal{P}^+_{\lambda,\Lambda}(D^2u) - b(x)\cdot Du = \beta\delta\lambda(1+|x|^2)^{-\frac{\delta}{2}-2} \geq 0 \text{ in } \mathbb{R}^N ,$$

while $\lim_{|x|\to\infty} b(x)\cdot x > \lambda(2-\beta)$.

Regarding the degenerate setting addressed in [**9**], and the sharpness of the conditions therein, similar counterexamples can be built for PDEs over Hörmander vector fields in the presence of a "fundamental solution". This issue has been recently investigated in [**10**].

Lastly, we show how to improve the condition in Theorem 4.1 when looking at supersolutions to $F = \mathcal{P}^-_{\lambda,\Lambda}$ instead of $F = \mathcal{P}^+_{\lambda,\Lambda}$ (as we did in the previous Remark 2.6). Here, the crucial point is that the difference between the supersolution and the Lyapunov function w, which is now a classical subsolution to $\mathcal{P}^-_{\lambda,\Lambda}(D^2u) - b\cdot Du$, turns out to be a supersolution of a problem involving $\mathcal{P}^+_{\lambda,\Lambda}$. That $\mathcal{P}^-$ switches into $\mathcal{P}^+$ is not a problem, since we still have the strong minimum principle and the comparison principle on bounded open sets.

THEOREM 4.7. *Let u be a viscosity supersolution bounded below to*

$$\mathcal{P}^-_{\lambda,\Lambda}(D^2u) - b(x)\cdot Du = A|Du|^\gamma \text{ in } \mathbb{R}^N,$$

and either

$$A > 0, \quad \text{and} \quad \gamma > 0$$

or

$$A \in \mathbb{R}, \quad \text{and} \quad \gamma = 2.$$

Assume also that b satisfies (4.2) *and*

$$\limsup_{|x|\to\infty}\{b(x)\cdot x\} < \Lambda - \lambda(N-1) . \tag{4.5}$$

Then, u is constant.

REMARK 4.8. Note that condition (4.5) is better than (4.3) since

$$\Lambda - \lambda(N-1) > \lambda - \Lambda(N-1) \iff N(\Lambda-\lambda) > 0 .$$

We further remark that in such case the Liouville property could have been deduced via the corresponding results for the Laplacian since $\mathcal{P}^-_{\lambda,\Lambda}(D^2u) \leq -\Lambda\Delta u$, where the condition would read as

$$\limsup_{|x|\to\infty}\{b(x)\cdot x\} < \Lambda(2-N) . \tag{4.6}$$

However, condition (4.5) is again better than (4.6), since

$$\Lambda - \lambda(N-1) > \Lambda(2-N) \iff (\Lambda-\lambda)(N-1) > 0 .$$

PROOF OF THEOREM 4.7. The proof is similar to Theorem 4.1. We only outline the main differences. The Liouville property for the viscosity inequality $\mathcal{P}^-_{\lambda,\Lambda}(D^2u) + b(x)\cdot Du \geq 0$ can be obtained exactly as in Lemma 4.2, using $w(x) = -\log|x|$ as a Lyapunov function (an explosive subsolution of $\mathcal{P}^-_{\lambda,\Lambda}(D^2u) - b\cdot Du$).

Note that for $\xi > 0$, by a standard test function argument together with the inequality $\mathcal{P}^+_{\lambda,\Lambda}(M+N) \geq \mathcal{P}^+_{\lambda,\Lambda}(M) + \mathcal{P}^-_{\lambda,\Lambda}(N)$, one verifies that $v_\xi = u - \xi w$ solves in viscosity sense

$$\mathcal{P}^+_{\lambda,\Lambda}(D^2 v_\xi) - b(x)\cdot Dv_\xi \geq 0 \text{ for every } x \text{ such that } |x| > \bar{R},$$

namely it is a supersolution of an equation involving the maximal operator $\mathcal{P}^+$ (and not the minimal operator $\mathcal{P}^-$). Still, one has the strong minimum principle and the comparison principle on the annuli to achieve the result.

In the case $H(Du) = -|Du|^2$, one argues as in Theorem 4.1, now using the inequality $\mathcal{P}^-_{\lambda,\Lambda}(M+N) \geq \mathcal{P}^-_{\lambda,\Lambda}(M)+\mathcal{P}^-_{\lambda,\Lambda}(N)$ to find that for $v(x) = \lambda(1-e^{-\frac{1}{\lambda}u(x)})$ and smooth u

$$\begin{aligned}
\mathcal{P}^-_{\lambda,\Lambda}(D^2v) - b(x)\cdot Dv &= \mathcal{P}^-_{\lambda,\Lambda}\left(e^{-\frac{1}{\lambda}u}D^2u - \frac{1}{\lambda}Du\otimes Du\, e^{-\frac{1}{\lambda}u}\right) - b(x)\cdot Du\, e^{-\frac{1}{\lambda}u} \\
&\geq \mathcal{P}^-_{\lambda,\Lambda}(D^2u)e^{-\frac{1}{\lambda}u} + \mathcal{P}^-_{\lambda,\Lambda}\left(-\frac{1}{\lambda}Du\otimes Du\, e^{-\frac{1}{\lambda}u}\right) - b(x)\cdot Du\, e^{-\frac{1}{\lambda}u} \\
&= \mathcal{P}^-_{\lambda,\Lambda}(D^2u)e^{-\frac{1}{\lambda}u} - \frac{1}{\lambda}\mathcal{P}^+_{\lambda,\Lambda}\left(Du\otimes Du\right)e^{-\frac{1}{\lambda}u} - b(x)\cdot Du\, e^{-\frac{1}{\lambda}u} \\
&= (\mathcal{P}^-_{\lambda,\Lambda}(D^2u) + |Du|^2 - b(x)\cdot Du)e^{-\frac{1}{\lambda}u} \geq 0\,.
\end{aligned}$$

□

REMARK 4.9. As a final suggestion, following [**27**, Theorem 1.1], a Liouville result as in Theorems 4.1 and 4.7 can be obtained for the problem

$$\mathcal{P}^+_{\lambda,\Lambda}(D^2u) - b(x)\cdot Du + u^q|Du|^2 \geq 0 \text{ in } \mathbb{R}^N\ , q\geq 0$$

via the transformation

$$v(x) = \int_0^{u(x)} e^{-\frac{s^{q+1}}{(q+1)\lambda}}\,ds\,.$$

Indeed, assuming that u is a smooth function, standard calculations lead to

$$Dv = e^{-\frac{u^{q+1}}{\lambda(q+1)}}Du\,;$$

$$D^2v = -e^{-\frac{u^{q+1}}{\lambda(q+1)}}\frac{u^q}{\lambda}Du\otimes Du + e^{-\frac{u^{q+1}}{\lambda(q+1)}}D^2u\,.$$

Then,

$$\begin{aligned}
\mathcal{P}^+_{\lambda,\Lambda}(D^2v) - b(x)\cdot Dv &= \mathcal{P}^+_{\lambda,\Lambda}\left(-e^{-\frac{u^{q+1}}{\lambda(q+1)}}\frac{u^q}{\lambda}Du\otimes Du + e^{-\frac{u^{q+1}}{\lambda(q+1)}}D^2u\right) \\
&\quad - e^{-\frac{u^{q+1}}{\lambda(q+1)}}b(x)\cdot Du \\
\geq \mathcal{P}^-_{\lambda,\Lambda}&\left(-e^{-\frac{u^{q+1}}{\lambda(q+1)}}\frac{u^q}{\lambda}Du\otimes Du\right) + \mathcal{P}^+_{\lambda,\Lambda}\left(e^{-\frac{u^{q+1}}{\lambda(q+1)}}D^2u\right) - e^{-\frac{u^{q+1}}{\lambda(q+1)}}b(x)\cdot Du \\
&= e^{-\frac{u^{q+1}}{\lambda(q+1)}}\left(u^q|Du|^2 + \mathcal{P}^+_{\lambda,\Lambda}(D^2u) - b(x)\cdot Du\right) \geq 0\,,
\end{aligned}$$

so that the Liouville property for u boils down to the one for v.

A similar transformation can be applied to problems driven by the singular operator

$$F(Du, D^2u) = |Du|^{m-2}\mathcal{P}^+_{\lambda,\Lambda}(D^2u),$$

$m > 1$, and nonlinearities of the form $H(u, Du) = u^q|Du|^\gamma$ with $m = \gamma$ and any $q \geq 0$, as in [**27**] for the case of the m-Laplacian.

5. Further remarks

We now discuss how to derive the Liouville property in some special situations.

REMARK 5.1. It is immediate to conclude that all the Liouville-type results obtained above lead to new Liouville properties for supersolutions to equations driven by non-divergence operators of the form

$$-\mathrm{Tr}(A(x,u,Du)D^2u) \geq H_i(x,u,Du) \text{ in } \mathbb{R}^N ,$$

where $\lambda I_N \leq A \leq \Lambda I_N$ (see for example Remark 5.6 below). We emphasize that the literature on Liouville properties for such linear operators is not as wide as the one on constant coefficient equations. See for example [**51**] for some results on problems with zero-th order terms. All the known properties in the non-divergence setting with the presence of gradient dependent nonlinearities have been actually obtained from those on fully nonlinear problems.

REMARK 5.2. In Sections 2 and 3 we analyzed Liouville-type results for $\beta > 2$ and $N > 2$. The cases $N = 1$ (that corresponds to $\beta \leq 2$) and $N = 2$ are in general simpler to address, as it happens for quasi-linear problems driven by the Laplacian, since the Liouville properties follow from the corresponding ones of superharmonic functions.

First, when $\beta \leq 2$, the Liouville properties for (classical) solutions of $F(x, D^2u) \geq H_i(x,u,Du)$ follow from the fact that there are no non-nonstant positive concave functions on the real line, since when $F(x,0) = 0$ and F is uniformly elliptic, it holds $\mathcal{P}^+_{\lambda,\Lambda}(u'') \geq F(x,u'')$ and so u solves $\mathcal{P}^+_{\lambda,\Lambda}(u'') \geq 0$ with

$$\mathcal{P}^+_{\lambda,\Lambda}(s) = \begin{cases} -\lambda s \text{ if } s \geq 0, \\ -\Lambda s \text{ if } s < 0. \end{cases}$$

When instead $N = 2$ and $F = \mathcal{P}^-_{\lambda,\Lambda}$, the result of Theorem 2.1 follows from the fact that the supersolutions to $F \geq H$ are superharmonic, so u is constant by e.g. [**9**, Theorem 2.1] or [**74**, Theorem 29 Ch. 2], and it holds for any $q, \gamma \geq 0$. Similarly, Conjecture 3.3 for $F = \mathcal{P}^-_{\lambda,\Lambda}$ is true by the same argument for any $q, \gamma \geq 0$ in $\mathbb{R}^2$. The case of arbitrary fully nonlinear operators in the plane (and in particular when $F = \mathcal{P}^+_{\lambda,\Lambda}$) is more delicate, since functions are no longer superharmonic in $\mathbb{R}^2$. If one exploits the comparison with the maximal operator, the condition ensuring the Liouville property becomes $N \leq \frac{\lambda}{\Lambda} + 1$ (see (4.3) with $b = 0$), which rules out the case $N = 2$. Therefore, even in the non-divergent case, the Liouville property in the plane is in general lost if no further conditions are imposed. Indeed, if one considers the equation $-\mathrm{Tr}(AD^2u) \geq H$, since $H \geq 0$, it follows that u is a solution to $-\mathrm{Tr}(AD^2u) \geq 0$ in $\mathbb{R}^2$, and a Liouville property could be obtained under further assumptions at infinity of the diffusion matrix, cf e.g. [**51**]. For instance, one can apply [**9**, Theorem 2.1] provided there exists a Lyapunov function for $-\mathrm{Tr}(A(x)D^2w) = 0$ in $\mathbb{R}^2$, $\lambda I_N \leq A \leq \Lambda I_N$. Standard calculations for a radial function $w(x) = f(|x|)$ lead to

$$-\mathrm{Tr}(A(x)D^2w) = -\frac{\mathrm{Tr}(Ax \otimes x)}{|x|^2}\left(f''(|x|) + \frac{\Psi_a(x) - 1}{|x|}f'(|x|)\right) ,$$

where $\Psi_a(x) = \frac{\mathrm{Tr}(A)}{\frac{\mathrm{Tr}(Ax\otimes x)}{|x|^2}}$ is the so-called effective dimension of the linear operator, cf [**61**, p.518], which is now the quantity responsible for the qualitative behavior of the linear operator. Therefore, if one takes $w(x) = -\log|x|$, one has

$$-\mathrm{Tr}(A(x)D^2w) = \frac{\mathrm{Tr}(Ax \otimes x)}{|x|^2}\left(\frac{\Psi_a(x) - 2}{|x|^2}\right).$$

Note that for $A = I_N$ it follows $\Psi_a(x) = N$, while Ψ_a stands for the linear counterpart of the intrinsic dimension β of the Pucci's maximal operator. Hence, when $A = I_2$ it follows that w is a subsolution in the plane, and the Liouville property holds for the Laplace equation in $\mathbb{R}^2$, cf [**9**, Remark 2.2]. For general uniformly elliptic operators in non-divergence form, heuristically, if the matrix of the coefficients approaches to a constant matrix for large $|x|$, the value responsible for the Liouville properties of the non-divergent equation would be the same as the one for the Laplacian.

REMARK 5.3. Owing to the same techniques and with appropriate modifications, the symmetric Liouville property for *subsolutions bounded above* to the viscosity inequality

$$F(x, D^2u) \leq H_i(x, u, Du) \quad \text{in } \mathbb{R}^N$$

can be obtained via the analysis of the partial differential inequality

$$\mathcal{P}^-_{\lambda,\Lambda}(D^2u) \leq H_i(x, u, Du) \quad \text{in } \mathbb{R}^N .$$

REMARK 5.4. We have discussed in this paper the Liouville properties for C-viscosity inequalities involving fully nonlinear operators. As far as we know, the analysis of similar properties in the framework of L^p-viscosity solutions is widely open. However, some results can be obtained in a straightforward manner from the linear theory. For example, if one looks at nonpositive (and not nonnegative!) L^p-viscosity solutions to the inequality $\mathcal{P}^-_{\lambda,\Lambda}(D^2u) \geq |Du|^\gamma$, it is possible to prove that u is a constant for every $\gamma > 0$ by requiring also $u \in L^p(\mathbb{R}^N)$, $1 < p < \infty$. Indeed, u solves also the inequality $-\Lambda\Delta u \geq |Du|^\gamma$ and in particular u is superharmonic. Then $v = -u$ is subharmonic and nonnegative. Combining the mean-value property and the Hölder's inequality one concludes that v is a constant, and so is u.

REMARK 5.5. It is well-known that the classical Liouville property fails for nonhomogeneous problems, a simple example being the function $u(x) = |x|^2$, which is a nonnegative and non-constant solution to $-\Delta u = k$ in $\mathbb{R}^N$ with $k = -2N$. Therefore, if one avoids the restriction $F(x, 0) = 0$, and the equation is (u, Du)-independent, it is possible to formulate and expect different Liouville properties. For instance, if one imposes the condition $F(x, 0) = k$ for some constant $k \in \mathbb{R}$ in the fully nonlinear case, one can expect to prove that u is a quadratic polynomial instead of a constant. An instance of such a Liouville phenomenon appears for the Monge-Ampère equation as a consequence of second order bounds, cf [**66**, Theorem 1.7.2] or [**64**, Remark 4], the so-called Jörgen-Calabi-Pogorelov Theorem, the Special Lagrangian equation [**66**, Theorem 1.7.3] and other Hessian equations, as discussed in e.g. [**65**, Corollary 1.2] and [**66**]. We conclude by saying that in the case of solutions to Hessian equations without x-dependence, a related assumption is crucial. For instance, when looking at solutions (not super- or subsolutions) to $F(D^2u) = 0$ in $\mathbb{R}^N$, the assumption $F(0) = 0$ is essential to deduce the Liouville property when $N \geq 5$, cf [**66**, Section 1.7.1]. However, it was conjectured, see [**66**, Conjecture 1.7.1], that such a restriction can be dropped in the lower dimensional cases $N \leq 4$.

REMARK 5.6. The above results involving fully nonlinear uniformly elliptic equations can be extended to problems driven by the normalized p-Laplacian: for $p \in (1, \infty)$, the operator

$$-\widetilde{\Delta}_p u = -\frac{1}{p}|Du|^{2-p}\mathrm{div}(|Du|^{p-2}Du) = -\frac{1}{p}\Delta u - \frac{p-2}{p}\frac{\Delta_\infty u}{|Du|^2},$$

can be indeed compared with Pucci's extremal operators. Unlike the p-Laplacian, we can write the above operator in non-divergence form as

$$-\widetilde{\Delta}_p u = -\mathrm{Tr}(A(Du)D^2 u)$$

where

$$A(Du) = \frac{1}{p}\left(I_N + (p-2)\frac{Du \otimes Du}{|Du|^2}\right),$$

which is positive definite, and bounded for every $p \in (1, \infty)$. Moreover, its eigenvalues are bounded below by $\lambda = \min\left\{\frac{1}{p}, \frac{p-1}{p}\right\}$ and $\Lambda = \max\left\{\frac{1}{p}, \frac{p-1}{p}\right\}$. We then have

$$\mathcal{P}^-_{\lambda,\Lambda}(D^2u) \le -\widetilde{\Delta}_p u \le \mathcal{P}^+_{\lambda,\Lambda}(D^2u)$$

whenever $\lambda = \min\left\{\frac{1}{p}, \frac{p-1}{p}\right\}$ and $\Lambda = \max\left\{\frac{1}{p}, \frac{p-1}{p}\right\}$. Therefore, we can state the following consequences of Theorems 2.1 and 4.1.

COROLLARY 5.7. *Any nonnegative viscosity solution to*

$$-\widetilde{\Delta}_p u \ge u^q + |Du|^\gamma \text{ in } \mathbb{R}^N,$$

with $p > 2$, $1 < \gamma \le \frac{(p-1)(N-1)+1}{(p-1)(N-1)}$ *and* $q > \frac{(p-1)(N-1)+1}{(p-1)(N-1)-1}$ *is constant. Instead, when* $1 < p < 2$ *the same assertion holds for* $1 < \gamma \le \frac{N+p-2}{N-1}$ *and* $q > \frac{N+p-2}{N-p}$.

COROLLARY 5.8. *Let u be a viscosity solution bounded from below to*

$$-\widetilde{\Delta}_p u + b(x)\cdot Du \ge A|Du|^\gamma \qquad \text{in } \mathbb{R}^N,$$

and either

$$A > 0, \quad \text{and} \quad \gamma > 0$$

or

$$A \in \mathbb{R}, \quad \text{and} \quad \gamma = 2.$$

Assume that b satisfies (4.2) *and*

$$\limsup_{|x|\to\infty} b(x)\cdot x < \begin{cases} 1 - \frac{N(p-1)}{p} & \text{if } p > 2 \\ 1 - \frac{N}{p} & \text{if } 1 < p \le 2. \end{cases} \tag{5.1}$$

Then, u is constant.

We finally mention that, for nonlinear problems considered above, one could address different forms of the Liouville property. The following questions are inspired by the works by V. Kurta (see e.g. [**47**] and the references therein).

OPEN PROBLEM 5.9 (Liouville comparison principle). It is known that if u, v solve in weak sense the inequality

$$-\Delta u - |u|^{q-1}u \le -\Delta v - |v|^{q-1}v \quad \text{in } \mathbb{R}^N, u \le v \text{ in } \mathbb{R}^N$$

and $1 < q \le \frac{N}{N-2}$, then $u \equiv v$ in $\mathbb{R}^N$, without any assumption on the behavior at infinity of the couple (u, v). Note that, taking $u = 0$ in the previous statement, one has exactly the Liouville property discussed in the previous sections. Does the result extend to PDEs involving superlinear gradient terms and/or involving fully nonlinear operators?

A preliminary answer for fully nonlinear rotationally invariant operators perturbed with zero-th order terms with superlinear power growth is provided by the following result, which is based on the so-called transitivity of the viscosity inequalities.

THEOREM 5.10. *Let* $1 \le q \le \frac{\beta}{\beta-2}$ *and* $u, v \in C(\mathbb{R}^N)$ *be respectively a viscosity sub- and supersolution to* $\mathcal{P}^+_{\lambda,\Lambda}(D^2 z) - |z|^{q-1}z = 0$ *in* $\mathbb{R}^N$ *satisfying* $u \le v$ *in* $\mathbb{R}^N$. *Then,* $u \equiv v$ *in* $\mathbb{R}^N$.

PROOF. We first observe that if $u \le v$ and they touch at some interior point x_0, then $u \equiv v$ in $\mathbb{R}^N$ by the strong comparison principle for Pucci's extremal operators, see e.g. [**44**]. Thus, we may assume $u < v$ in $\mathbb{R}^N$. By the transitivity of viscosity inequalities Lemma 4.3, and the algebraic inequality

$$(|v|^{q-1}v - |u|^{q-1}u)(v-u) \ge 2^{1-q}|v-u|^{q+1} ,$$

it follows that $w = v - u \ge 0$ solves in viscosity sense

$$\mathcal{P}^+_{\lambda,\Lambda}(D^2 w) \ge 2^{1-q} w^q \quad \text{in } \mathbb{R}^N .$$

Since $u \le v$, we have

$$w \ge 0 , \quad \mathcal{P}^+_{\lambda,\Lambda}(D^2 w) \ge 0 \quad \text{in } \mathbb{R}^N .$$

Then, one can proceed as in [**32**, Theorem 4.1] to conclude that $w \equiv 0$ in $\mathbb{R}^N$, i.e. $u \equiv v$ in $\mathbb{R}^N$. □

OPEN PROBLEM 5.11. When the Liouville property fails, it is well-known that for problems driven by the Laplacian or the p-Laplacian it is possible to establish the sharp distance at infinity from the non-constant supersolution to the constant one, see e.g. [**53**, Theorem 1.2], [**52**]. Does this quantitative property extend to the fully nonlinear setting?

We conclude with an open problem for truncated Laplacians, as studied in e.g. [**21**].

OPEN PROBLEM 5.12. Do the results of the previous sections extend to the problem $-\mathcal{P}^+_k(D^2u) \ge H_i(u, Du)$, $u \ge 0$ in $\mathbb{R}^N$, where $\mathcal{P}^+_k(X) = \sum_{i=N-k+1}^N e_i(X)$, for $2 < k < N$?

6. The Liouville property for solutions

In this final section, we collect some remarks on the Liouville property for *solutions* to elliptic PDEs involving super-linear, first-order perturbations. Our starting point is the well-known result stating that any classical solution to

$$-\Delta u + |Du|^\gamma = 0 \text{ in } \mathbb{R}^N , \gamma > 1$$

must be constant, see [**68**, p.67] or [**56**, Corollary IV]. Remarkably, no one-side bounds on u are required. Such a strong result, which (partially) extends to quasi-linear problems, has important consequences in the study of the regularity and the qualitative behavior of solutions. In this direction, the literature is extensive, and we refer to [**17, 27, 81**] for further references.

In what follows, we discuss some open questions (to our knowledge) regarding generalizations of this result to equations involving more general operators, with particular emphasis to subelliptic and nonlocal problems. We stress that all the aforementioned results have been obtained through (local) gradient estimates, generally accomplished via refinements of the Bernstein method (see also the more recent works [**17**,**38**]). However, such gradient bounds seem to be difficult to be reproduced both in the subelliptic and the nonlocal frameworks due to the structure of the diffusion operator.

6.1. The subelliptic case. We consider here the Liouville property for solutions to problems structured over Hörmander vector fields, with *quadratic* gradient terms (that is, the so-called natural gradient growth).

Consider a family $\mathcal{X} = \{X_1, \dots, X_m\}$, $m \leq N$, of linearly independent smooth vector fields on $\mathbb{R}^N$, $N \geq 3$, having the following properties:

(i) X_i's are δ_λ homogeneous of degree one with respect to a family of non-isotropic dilations defined as
$$\delta_\lambda : \mathbb{R}^N \to \mathbb{R}^N \ , \delta_\lambda(x) = (\lambda^{\sigma_1} x_1, \dots, \lambda^{\sigma_N} x_N)$$
where $1 = \sigma_1 \leq \dots \leq \sigma_N$ are positive integers, i.e. $X_i(\varphi(\delta_\lambda(x))) = \lambda(X_i\varphi)(\delta_\lambda(x))$ for every $\varphi \in C^\infty(\mathbb{R}^N)$, $x \in \mathbb{R}^N$, $\lambda > 0$;

(ii) The system $\mathcal{X}$ satisfies the Hörmander condition, see [**22**] and the references therein.

We denote by $D_{\mathcal{X}} u = (X_1 u, \dots, X_m u) \in \mathbb{R}^m$, $(D^2_{\mathcal{X}} u)^* = \frac{X_i X_j u + X_j X_i u}{2} \in \mathrm{Sym}_m$ respectively the horizontal gradient and Hessian of the unknown u, and $\Delta_{\mathcal{X}} u = \mathrm{Tr}((D^2_{\mathcal{X}} u)^*) = \sum_{i=1}^m X_i^2 u$. We recall that, for instance, assumptions (i)-(ii) are satisfied by the vector fields generating stratified Lie groups and Grushin-type geometries in $\mathbb{R}^N$, $N \geq 3$. We premise the following

LEMMA 6.1 (Degenerate Hopf-Cole transformation). *Let u be a classical solution to $-\Delta_{\mathcal{X}} u + b(x) \cdot D_{\mathcal{X}} u + |D_{\mathcal{X}} u|^2 = f(x)$. Then $v = e^{-u}$ solves*
$$-\Delta_{\mathcal{X}} v + b(x) \cdot D_{\mathcal{X}} v + f(x) v = 0 \ .$$

PROOF. Using standard calculus rules in the subelliptic setting, cf [**22**], we write
$$D_{\mathcal{X}} v = -v D_{\mathcal{X}} u$$
$$D^2_{\mathcal{X}} v = v D_{\mathcal{X}} u \otimes D_{\mathcal{X}} u - v D^2_{\mathcal{X}} u \implies \Delta_{\mathcal{X}} v = \mathrm{Tr}(D^2_{\mathcal{X}} v) = v(|D_{\mathcal{X}} u|^2 - \Delta_{\mathcal{X}} u) \ .$$
Then
$$\begin{aligned} -\Delta_{\mathcal{X}} v + b(x) \cdot D_{\mathcal{X}} v + f(x) v &= -v(|D_{\mathcal{X}} u|^2 - \Delta_{\mathcal{X}} u) - v b(x) \cdot D_{\mathcal{X}} u + f(x) v \\ &= v[\Delta_{\mathcal{X}} u - |D_{\mathcal{X}} u|^2 - b(x) \cdot D_{\mathcal{X}} u + f(x)] = 0 \ . \end{aligned}$$
□

As a simple consequence one can deduce the following result, that also gives an alternative proof to [**68**, p.67], [**56**, Corollary IV] in the quadratic case $\gamma = 2$ for Euclidean vector fields.

THEOREM 6.2. *Assume (i)-(ii). Let u be a classical solution of*
$$-\Delta_{\mathcal{X}} u + |D_{\mathcal{X}} u|^2 = 0 \text{ in } \mathbb{R}^N \ . \tag{6.1}$$
Then, u is constant.

PROOF. We use the Hopf-Cole transform in Lemma 6.1 to show that if u solves in classical sense (6.1), then $v = e^{-u} > 0$ is a positive solution to

$$-\Delta_{\mathcal{X}} v = 0 \text{ in } \mathbb{R}^N .$$

We then apply the Liouville property in [**48**, Corollary 8.3], [**13**, Proposition 5.5] (which are obtained as a consequence of Harnack inequalities) to conclude that v is constant since it is one-side bounded. Then, also u is constant. □

REMARK 6.3. A similar property holds for solutions to nonlinear problems driven by the stationary Kolmogorov operator

$$v \cdot D_x u - \Delta_v u + |D_v u|^2 = 0 \text{ in } \mathbb{R}^{2N} , (x, v) \in \mathbb{R}^N \times \mathbb{R}^N .$$

Indeed, $w = e^{-u} > 0$ solves $v \cdot D_x w - \Delta_v w = 0$ in $\mathbb{R}^{2N}$ and w is constant by the Liouville property obtained in [**48**] via the Harnack inequality, since w is one-side bounded.

While the proof in the special quadratic case is rather easy (basically it requires a linear result only), the non-quadratic case appears to be challenging. Indeed, classical results [**68**, p.67] or [**56**, Corollary IV] are consequence of gradient bounds (obtained via the so-called Bernstein method). It is not clear at this stage how to prove analogous bounds in the subelliptic framework.

We now give some comments on possible generalizations to the fully nonlinear framework. The problem seems to be open even in the Euclidean setting.

OPEN PROBLEM 6.4. Let $\gamma > 1$ and u be a viscosity solution to

$$\mathcal{P}^{\pm}_{\lambda,\Lambda}(D^2 u) + |Du|^\gamma = 0 \text{ in } \mathbb{R}^N .$$

Can we conclude that u is a constant (without assuming any bound on u)?

We emphasize that even in the quadratic case $\gamma = 2$, the result cannot be deduced through the Hopf-Cole transformation, since such change of variable leads to an inequality (as in Theorem 4.1), rather than to an equality. Still, we mention that some results in these direction recently appeared in [**39**].

More generally, having in mind the works [**18, 38, 56, 68**], one could formulate the

OPEN PROBLEM 6.5. Let $F((D^2_{\mathcal{X}} u)^*) = \mathcal{P}^{\pm}_{\lambda,\Lambda}((D^2_{\mathcal{X}} u)^*)$ or the p-sub-Laplacian $-\Delta_{p,\mathcal{X}} u = -\mathrm{div}_{\mathcal{X}}(|D_{\mathcal{X}} u|^{p-2} D_{\mathcal{X}} u)$, $\mathrm{div}_{\mathcal{X}}(\Phi(x)) = \sum_{i=1}^m X_i \Phi$, $\Phi : \mathbb{R}^N \to \mathbb{R}^m$. Does the one (or two) side Liouville property for solutions to

$$F((D^2_{\mathcal{X}} u)^*) \pm u^q |D_{\mathcal{X}} u|^\gamma = 0 \text{ in } \mathbb{R}^N , q > 0 , \gamma > 1 ,$$

hold? Does the property hold without bounds on the solution as in [**56, 68**]?

6.2. Nonlocal problems. The derivation of Liouville properties for solutions to equations driven by nonlocal operators, like the fractional Laplacian $(-\Delta)^s$, $s \in (0, 1)$, perturbed by gradient terms with superlinear nature, seems to be also a nontrivial issue. We mention the following

OPEN PROBLEM 6.6. Does the Liouville property hold for solutions to

$$\mathcal{L}_s u \pm |Du|^\gamma = 0 \text{ in } \mathbb{R}^N ,$$

where $\mathcal{L}_s = -\Delta + (-\Delta)^s$ or $\mathcal{L}_s = (-\Delta)^s$, $s \in (0, 1)$, $\gamma > 0$?

As in the subelliptic setting, the derivation of gradient estimates via Bernstein-type arguments seems to be not feasible (or, at least, not straightforward), and a different procedure might be required. In this direction, we mention some recent results obtained in [**3**] for linear equations, which combine PDE and group theory techniques.

References

[1] S. Alarcón, J. García-Melián, and A. Quaas, *Nonexistence of positive supersolutions to some nonlinear elliptic problems* (English, with English and French summaries), J. Math. Pures Appl. (9) **99** (2013), no. 5, 618–634, DOI 10.1016/j.matpur.2012.10.001. MR3039209

[2] Salomón Alarcón, Jorge García-Melián, and Alexander Quaas, *Optimal Liouville theorems for supersolutions of elliptic equations with the Laplacian*, Ann. Sc. Norm. Super. Pisa Cl. Sci. (5) **16** (2016), no. 1, 129–158. MR3524667

[3] Nathaël Alibaud, Félix del Teso, Jørgen Endal, and Espen R. Jakobsen, *The Liouville theorem and linear operators satisfying the maximum principle* (English, with English and French summaries), J. Math. Pures Appl. (9) **142** (2020), 229–242, DOI 10.1016/j.matpur.2020.08.008. MR4149690

[4] Scott N. Armstrong and Boyan Sirakov, *Nonexistence of positive supersolutions of elliptic equations via the maximum principle*, Comm. Partial Differential Equations **36** (2011), no. 11, 2011–2047, DOI 10.1080/03605302.2010.534523. MR2846170

[5] Scott N. Armstrong and Boyan Sirakov, *Sharp Liouville results for fully nonlinear equations with power-growth nonlinearities*, Ann. Sc. Norm. Super. Pisa Cl. Sci. (5) **10** (2011), no. 3, 711–728. MR2905384

[6] Scott N. Armstrong, Boyan Sirakov, and Charles K. Smart, *Fundamental solutions of homogeneous fully nonlinear elliptic equations*, Comm. Pure Appl. Math. **64** (2011), no. 6, 737–777, DOI 10.1002/cpa.20360. MR2663711

[7] Martino Bardi and Annalisa Cesaroni, *Liouville properties and critical value of fully nonlinear elliptic operators*, J. Differential Equations **261** (2016), no. 7, 3775–3799, DOI 10.1016/j.jde.2016.06.006. MR3532054

[8] Martino Bardi and Alessandro Goffi, *New strong maximum and comparison principles for fully nonlinear degenerate elliptic PDEs*, Calc. Var. Partial Differential Equations **58** (2019), no. 6, Paper No. 184, 20, DOI 10.1007/s00526-019-1620-2. MR4018316

[9] Martino Bardi and Alessandro Goffi, *Liouville results for fully nonlinear equations modeled on Hörmander vector fields: I. The Heisenberg group*, Math. Ann., doi:10.1007/s00208-020-02118-x, 2020.

[10] Martino Bardi and Alessandro Goffi, *Liouville results for fully nonlinear equations modeled on Hörmander vector fields: II. Carnot groups and Grushin geometries*, `arXiv:2109.11331`, 2021.

[11] Begoña Barrios and Leandro M. Del Pezzo, *Study of the existence of supersolutions for nonlocal equations with gradient terms*, Milan J. Math. **88** (2020), no. 2, 267–294, DOI 10.1007/s00032-020-00314-7. MR4182074

[12] H. Berestycki, I. Capuzzo-Dolcetta, and L. Nirenberg, *Superlinear indefinite elliptic problems and nonlinear Liouville theorems*, Topol. Methods Nonlinear Anal. **4** (1994), no. 1, 59–78, DOI 10.12775/TMNA.1994.023. MR1321809

[13] Stefano Biagi and Ermanno Lanconelli, *Large sets at infinity and maximum principle on unbounded domains for a class of sub-elliptic operators*, J. Differential Equations **269** (2020), no. 11, 9680–9719, DOI 10.1016/j.jde.2020.06.060. MR4120381

[14] Davide Bianchi, Stefano Pigola, and Alberto G. Setti, *Qualitative properties of bounded subsolutions of nonlinear PDEs* (English, with English and French summaries), J. Math. Pures Appl. (9) **144** (2020), 137–163, DOI 10.1016/j.matpur.2020.11.003. MR4175447

[15] Bruno Bianchini, Luciano Mari, Patrizia Pucci, and Marco Rigoli, *Geometric analysis of quasilinear inequalities on complete manifolds—maximum and compact support principles and detours on manifolds*, Frontiers in Mathematics, Birkhäuser/Springer, Cham, [2021] ©2021, DOI 10.1007/978-3-030-62704-1. MR4241012

[16] Marie-Françoise Bidaut-Véron, *Local and global behavior of solutions of quasilinear equations of Emden-Fowler type*, Arch. Rational Mech. Anal. **107** (1989), no. 4, 293–324, DOI 10.1007/BF00251552. MR1004713

[17] Marie-Françoise Bidaut-Véron, *Liouville results and asymptotics of solutions of a quasilinear elliptic equation with supercritical source gradient term*, Adv. Nonlinear Stud. **21** (2021), no. 1, 57–76, DOI 10.1515/ans-2020-2109. MR4234084

[18] Marie-Françoise Bidaut-Véron, Marta García-Huidobro, and Laurent Véron, *Estimates of solutions of elliptic equations with a source reaction term involving the product of the function and its gradient*, Duke Math. J. **168** (2019), no. 8, 1487–1537, DOI 10.1215/00127094-2018-0067. MR3959864

[19] Isabeau Birindelli and Françoise Demengel, *Some Liouville theorems for the p-Laplacian*, Proceedings of the 2001 Luminy Conference on Quasilinear Elliptic and Parabolic Equations and System, Electron. J. Differ. Equ. Conf., vol. 8, Southwest Texas State Univ., San Marcos, TX, 2002, pp. 35–46. MR1990294

[20] Isabeau Birindelli and Françoise Demengel, *Comparison principle and Liouville type results for singular fully nonlinear operators* (English, with English and French summaries), Ann. Fac. Sci. Toulouse Math. (6) **13** (2004), no. 2, 261–287. MR2126744

[21] Isabeau Birindelli, Giulio Galise, and Fabiana Leoni, *Liouville theorems for a family of very degenerate elliptic nonlinear operators*, Nonlinear Anal. **161** (2017), 198–211, DOI 10.1016/j.na.2017.06.002. MR3673001

[22] A. Bonfiglioli, E. Lanconelli, and F. Uguzzoni, *Stratified Lie groups and potential theory for their sub-Laplacians*, Springer Monographs in Mathematics, Springer, Berlin, 2007. MR2363343

[23] M. Á. Burgos-Pérez, J. García-Melián, and A. Quaas, *Classification of supersolutions and Liouville theorems for some nonlinear elliptic problems*, Discrete Contin. Dyn. Syst. **36** (2016), no. 9, 4703–4721, DOI 10.3934/dcds.2016004. MR3541501

[24] Luis A. Caffarelli and Xavier Cabré, *Fully nonlinear elliptic equations*, American Mathematical Society Colloquium Publications, vol. 43, American Mathematical Society, Providence, RI, 1995, DOI 10.1090/coll/043. MR1351007

[25] I. Capuzzo Dolcetta and A. Cutrì, *Hadamard and Liouville type results for fully nonlinear partial differential inequalities*, Commun. Contemp. Math. **5** (2003), no. 3, 435–448, DOI 10.1142/S0219199703001014. MR1992357

[26] Gabriella Caristi and Enzo Mitidieri, *Nonexistence of positive solutions of quasilinear equations*, Adv. Differential Equations **2** (1997), no. 3, 319–359. MR1441847

[27] Caihong Chang, Bei Hu, and Zhengce Zhang, *Liouville-type theorems and existence of solutions for quasilinear elliptic equations with nonlinear gradient terms*, Nonlinear Anal. **220** (2022), Paper No. 112873, 29, DOI 10.1016/j.na.2022.112873. MR4400071

[28] Marco Cirant, *Stationary focusing mean-field games*, Comm. Partial Differential Equations **41** (2016), no. 8, 1324–1346, DOI 10.1080/03605302.2016.1192647. MR3532395

[29] Marco Cirant and Alessandro Goffi, *On the problem of maximal L^q-regularity for viscous Hamilton-Jacobi equations*, Arch. Ration. Mech. Anal. **240** (2021), no. 3, 1521–1534, DOI 10.1007/s00205-021-01641-8. MR4264951

[30] Tobias H. Colding and William P. Minicozzi II, *Liouville properties*, ICCM Not. **7** (2019), no. 1, 16–26, DOI 10.4310/ICCM.2019.v7.n1.a10. MR3960545

[31] Michael G. Crandall, Hitoshi Ishii, and Pierre-Louis Lions, *User's guide to viscosity solutions of second order partial differential equations*, Bull. Amer. Math. Soc. (N.S.) **27** (1992), no. 1, 1–67, DOI 10.1090/S0273-0979-1992-00266-5. MR1118699

[32] Alessandra Cutrì and Fabiana Leoni, *On the Liouville property for fully nonlinear equations* (English, with English and French summaries), Ann. Inst. H. Poincaré C Anal. Non Linéaire **17** (2000), no. 2, 219–245, DOI 10.1016/S0294-1449(00)00109-8. MR1753094

[33] Lorenzo D'Ambrosio and Enzo Mitidieri, *A priori estimates, positivity results, and nonexistence theorems for quasilinear degenerate elliptic inequalities*, Adv. Math. **224** (2010), no. 3, 967–1020, DOI 10.1016/j.aim.2009.12.017. MR2628800

[34] Patricio Felmer and Alexander Quaas, *Fundamental solutions and Liouville type theorems for nonlinear integral operators*, Adv. Math. **226** (2011), no. 3, 2712–2738, DOI 10.1016/j.aim.2010.09.023. MR2739791

[35] Patricio Felmer, Alexander Quaas, and Boyan Sirakov, *Solvability of nonlinear elliptic equations with gradient terms*, J. Differential Equations **254** (2013), no. 11, 4327–4346, DOI 10.1016/j.jde.2013.03.003. MR3035435

[36] Roberta Filippucci, *Nonexistence of positive weak solutions of elliptic inequalities*, Nonlinear Anal. **70** (2009), no. 8, 2903–2916, DOI 10.1016/j.na.2008.12.018. MR2509378

[37] Roberta Filippucci, *Nonexistence of nonnegative solutions of elliptic systems of divergence type*, J. Differential Equations **250** (2011), no. 1, 572–595, DOI 10.1016/j.jde.2010.09.028. MR2737854

[38] Roberta Filippucci, Patrizia Pucci, and Philippe Souplet, *A Liouville-type theorem for an elliptic equation with superquadratic growth in the gradient*, Adv. Nonlinear Stud. **20** (2020), no. 2, 245–251, DOI 10.1515/ans-2019-2070. MR4095468

[39] Rodrigo Fuentes and Alexander Quaas, *A note on one-dimensional symmetry for Hamilton-Jacobi equations with extremal Pucci operators and application to Bernstein type estimate*, NoDEA Nonlinear Differential Equations Appl. **29** (2022), no. 3, Paper No. 29, 22, DOI 10.1007/s00030-022-00761-8. MR4400169

[40] Basilis Gidas, *Symmetry properties and isolated singularities of positive solutions of nonlinear elliptic equations*, Nonlinear partial differential equations in engineering and applied science (Proc. Conf., Univ. Rhode Island, Kingston, R.I., 1979), Lecture Notes in Pure and Appl. Math., vol. 54, Dekker, New York, 1980, pp. 255–273. MR577096

[41] B. Gidas and J. Spruck, *A priori bounds for positive solutions of nonlinear elliptic equations*, Comm. Partial Differential Equations **6** (1981), no. 8, 883–901, DOI 10.1080/03605308108820196. MR619749

[42] David Gilbarg and Neil S. Trudinger, *Elliptic partial differential equations of second order*, 2nd ed., Grundlehren der mathematischen Wissenschaften [Fundamental Principles of Mathematical Sciences], vol. 224, Springer-Verlag, Berlin, 1983, DOI 10.1007/978-3-642-61798-0. MR737190

[43] Alessandro Goffi, *Some new Liovuille-type results for fully nonlinear PDEs on the Heisenberg group*, Nonlinear Anal. **200** (2020), 112013, 18, DOI 10.1016/j.na.2020.112013. MR4108439

[44] Alessandro Goffi and Francesco Pediconi, *A note on the strong maximum principle for fully nonlinear equations on Riemannian manifolds*, J. Geom. Anal. **31** (2021), no. 8, 8641–8665, DOI 10.1007/s12220-021-00607-2. MR4293943

[45] Alexander Grigor′yan, *Analytic and geometric background of recurrence and non-explosion of the Brownian motion on Riemannian manifolds*, Bull. Amer. Math. Soc. (N.S.) **36** (1999), no. 2, 135–249, DOI 10.1090/S0273-0979-99-00776-4. MR1659871

[46] Alexander Grigor'yan and Yuhua Sun, *On nonnegative solutions of the inequality $\Delta u + u^\sigma \leq 0$ on Riemannian manifolds*, Comm. Pure Appl. Math. **67** (2014), no. 8, 1336–1352, DOI 10.1002/cpa.21493. MR3225632

[47] Bernd Kawohl and Vasilii Kurta, *A Liouville comparison principle for solutions of singular quasilinear elliptic second-order partial differential inequalities*, Commun. Pure Appl. Anal. **10** (2011), no. 6, 1747–1762, DOI 10.3934/cpaa.2011.10.1747. MR2805337

[48] Alessia Elisabetta Kogoj and Ermanno Lanconelli, *An invariant Harnack inequality for a class of hypoelliptic ultraparabolic equations*, Mediterr. J. Math. **1** (2004), no. 1, 51–80, DOI 10.1007/s00009-004-0004-8. MR2088032

[49] Alessia E. Kogoj and Ermanno Lanconelli, *L^p-Liouville theorems for invariant partial differential operators in $\mathbb{R}^n$*, Nonlinear Anal. **121** (2015), 188–205, DOI 10.1016/j.na.2014.12.004. MR3348921

[50] Alessia E. Kogoj, Ermanno Lanconelli, and Enrico Priola, *Harnack inequality and Liouville-type theorems for Ornstein-Uhlenbeck and Kolmogorov operators*, Math. Eng. **2** (2020), no. 4, 680–697, DOI 10.3934/mine.2020031. MR4142365

[51] Vladimir Kondratiev, Vitali Liskevich, and Zeev Sobol, *Positive super-solutions to semi-linear second-order non-divergence type elliptic equations in exterior domains*, Trans. Amer. Math. Soc. **361** (2009), no. 2, 697–713, DOI 10.1090/S0002-9947-08-04453-X. MR2452821

[52] Vasilii V. Kurta, *On a Liouville phenomenon for entire weak supersolutions of elliptic partial differential equations* (English, with English and French summaries), Ann. Inst. H. Poincaré C Anal. Non Linéaire **23** (2006), no. 6, 839–848, DOI 10.1016/j.anihpc.2005.12.001. MR2271696

[53] Vasilii V. Kurta, *A Liouville comparison principle for solutions of semilinear elliptic partial differential inequalities*, Proc. Roy. Soc. Edinburgh Sect. A **138** (2008), no. 1, 139–155, DOI 10.1017/S0308210506000552. MR2388940

[54] E. M. Landis, *Second order equations of elliptic and parabolic type*, Translations of Mathematical Monographs, vol. 171, American Mathematical Society, Providence, RI, 1998. Translated from the 1971 Russian original by Tamara Rozhkovskaya; With a preface by Nina Ural′tseva, DOI 10.1090/mmono/171. MR1487894

[55] YanYan Li, *Some Liouville theorems and applications*, Perspectives in nonlinear partial differential equations, Contemp. Math., vol. 446, Amer. Math. Soc., Providence, RI, 2007, pp. 313–317, DOI 10.1090/conm/446/08638. MR2376666

[56] P.-L. Lions, *Quelques remarques sur les problèmes elliptiques quasilinéaires du second ordre* (French), J. Analyse Math. **45** (1985), 234–254, DOI 10.1007/BF02792551. MR833413

[57] Pierre-Louis Lions and Marek Musiela, *Ergodicity of diffusion processes*, Unpublished manuscript, 2002.

[58] Paola Mannucci, Claudio Marchi, and Nicoletta Tchou, *The ergodic problem for some subelliptic operators with unbounded coefficients*, NoDEA Nonlinear Differential Equations Appl. **23** (2016), no. 4, Art. 47, 26, DOI 10.1007/s00030-016-0401-2. MR3530936

[59] Luciano Mari and Leandro F. Pessoa, *Duality between Ahlfors-Liouville and Khas'minskii properties for non-linear equations*, Comm. Anal. Geom. **28** (2020), no. 2, 395–497, DOI 10.4310/CAG.2020.v28.n2.a6. MR4101343

[60] Luciano Mari and Daniele Valtorta, *On the equivalence of stochastic completeness and Liouville and Khas'minskii conditions in linear and nonlinear settings*, Trans. Amer. Math. Soc. **365** (2013), no. 9, 4699–4727, DOI 10.1090/S0002-9947-2013-05765-0. MR3066769

[61] Norman Meyers and James Serrin, *The exterior Dirichlet problem for second order elliptic partial differential equations*, J. Math. Mech. **9** (1960), 513–538, DOI 10.1512/iumj.1960.9.59029. MR0117421

[62] È. Mitidieri and S. I. Pokhozhaev, *A priori estimates and the absence of solutions of nonlinear partial differential equations and inequalities* (Russian, with English and Russian summaries), Tr. Mat. Inst. Steklova **234** (2001), 1–384; English transl., Proc. Steklov Inst. Math. **3(234)** (2001), 1–362. MR1879326

[63] Enzo Mitidieri and Stanislav I. Pohozaev, *Towards a unified approach to nonexistence of solutions for a class of differential inequalities*, Milan J. Math. **72** (2004), 129–162, DOI 10.1007/s00032-004-0032-7. MR2099130

[64] C. Mooney, *The Monge-Ampère equation*, Rend. Semin. Mat. Univ. Politec. Torino **76** (2018), no. 1, 93–113. MR3956535

[65] Connor Mooney and Ovidiu Savin, *Regularity results for the equation $u_{11}u_{22} = 1$*, Discrete Contin. Dyn. Syst. **39** (2019), no. 12, 6865–6876, DOI 10.3934/dcds.2019235. MR4026172

[66] Nikolai Nadirashvili, Vladimir Tkachev, and Serge Vlăduţ, *Nonlinear elliptic equations and nonassociative algebras*, Mathematical Surveys and Monographs, vol. 200, American Mathematical Society, Providence, RI, 2014, DOI 10.1090/surv/200. MR3243534

[67] Edward Nelson, *A proof of Liouville's theorem*, Proc. Amer. Math. Soc. **12** (1961), 995, DOI 10.2307/2034412. MR259149

[68] L. A. Peletier and J. Serrin, *Gradient bounds and Liouville theorems for quasilinear elliptic equations*, Ann. Scuola Norm. Sup. Pisa Cl. Sci. (4) **5** (1978), no. 1, 65–104. MR481493

[69] Stefano Pigola, Marco Rigoli, and Alberto G. Setti, *A remark on the maximum principle and stochastic completeness*, Proc. Amer. Math. Soc. **131** (2003), no. 4, 1283–1288, DOI 10.1090/S0002-9939-02-06672-8. MR1948121

[70] Stefano Pigola, Marco Rigoli, and Alberto G. Setti, *Some non-linear function theoretic properties of Riemannian manifolds*, Rev. Mat. Iberoam. **22** (2006), no. 3, 801–831. MR2320402

[71] Stefano Pigola, Marco Rigoli, and Alberto G. Setti, *Aspects of potential theory on manifolds, linear and non-linear*, Milan J. Math. **76** (2008), 229–256, DOI 10.1007/s00032-008-0084-1. MR2465992

[72] A. Porretta and E. Priola, *Global Lipschitz regularizing effects for linear and nonlinear parabolic equations* (English, with English and French summaries), J. Math. Pures Appl. (9) **100** (2013), no. 5, 633–686, DOI 10.1016/j.matpur.2013.01.016. MR3115828

[73] Enrico Priola and Jerzy Zabczyk, *Liouville theorems for non-local operators*, J. Funct. Anal. **216** (2004), no. 2, 455–490, DOI 10.1016/j.jfa.2004.04.001. MR2095690

[74] Murray H. Protter and Hans F. Weinberger, *Maximum principles in differential equations*, Springer-Verlag, New York, 1984. Corrected reprint of the 1967 original, DOI 10.1007/978-1-4612-5282-5. MR762825

[75] Alexander Quaas and Boyan Sirakov, *Existence and non-existence results for fully nonlinear elliptic systems*, Indiana Univ. Math. J. **58** (2009), no. 2, 751–788, DOI 10.1512/iumj.2009.58.3501. MR2514387

[76] Pavol Quittner and Philippe Souplet, *Superlinear parabolic problems*, Birkhäuser Advanced Texts: Basler Lehrbücher. [Birkhäuser Advanced Texts: Basel Textbooks], Birkhäuser/Springer, Cham, 2019. Blow-up, global existence and steady states; Second edition of [MR2346798], DOI 10.1007/978-3-030-18222-9. MR3967048

[77] Luca Rossi, *Non-existence of positive solutions of fully nonlinear elliptic equations in unbounded domains*, Commun. Pure Appl. Anal. **7** (2008), no. 1, 125–141, DOI 10.3934/cpaa.2008.7.125. MR2358359

[78] D. Ruiz, *A priori estimates and existence of positive solutions for strongly nonlinear problems*, J. Differential Equations **199** (2004), no. 1, 96–114, DOI 10.1016/j.jde.2003.10.021. MR2041513

[79] James Serrin, *Entire solutions of nonlinear Poisson equations*, Proc. London Math. Soc. (3) **24** (1972), 348–366, DOI 10.1112/plms/s3-24.2.348. MR289961

[80] Yuhua Sun and Fanheng Xu, *Liouville's theorem to quasilinear differential inequalities involving gradient nonlinearity term on manifolds*, `arXiv:2102.02073`, 2021.

[81] Laurent Véron, *Local and global aspects of quasilinear degenerate elliptic equations*, World Scientific Publishing Co. Pte. Ltd., Hackensack, NJ, 2017. Quasilinear elliptic singular problems, DOI 10.1142/9850. MR3676369

Dipartimento di Matematica "Tullio Levi-Civita", Università degli Studi di Padova, via Trieste 63, 35121 Padova, Italy

Email address: `cirant@math.unipd.it`

Dipartimento di Scienze di Base ed Applicate per l'Ingegneria, Sapienza Università di Roma, Via Antonio Scarpa 10, 00161 Roma, Italy

Email address: `alessandro.goffi@uniroma1.it`

Contemporary Mathematics
Volume **781**, 2023
https://doi.org/10.1090/conm/781/15708

Regularity of the solution to the principal-agent problem

Shibing Chen

ABSTRACT. In this paper we prove the C^1 regularity of solution to the principal-agent problem with general preference function satisfying some structural conditions.

1. Introduction

One typical example of the principal-agent problem is that a monopolist wants to market automobiles ($y \in Y$) to a population of potential buyers ($x \in X$), with the following known informations: preference function $b(x, y)$ that measures the preference of a buyer x to the car y, the density of different types of buyers in the population $d\mu(x)$ and the cost $c(y)$ for manufacturing a car of type y. In order to maximize her profit, the principal needs to decide what kind of cars to manufacture and the price of each type of car. Recently Figalli, Kim and McCann [**6**] identified the conditions that ensure the existence and uniqueness of the strategy that the principal can use for maximizing her profit, they also discussed various interesting phenomena associated to the optimal strategy.

For some important special preference functions, the problem was studied by Wilson [**8**], Armstrong [**1**], and Rochet and Choné [**4**]. Later, for general preference function Carlier [**3**] reformulated it as a minimization problem over the space of b-convex functions, however for general $b(x, y)$ the space of admissible functions is generally not convex, which is the main reason that except existence he can not deduce other important properties such as the uniqueness of the strategy. By adapting a strengthened version of the so called MTW condition in optimal transportation [**7**] together with the conditions bi-twist and bi-convexity which were also previously used in the regularity theory of optimal transportation, Figalli, Kim and McCann successfully established various convexity properties in the principal-agent problem, and those convexity properties enable them to prove the uniqueness and some other important properties of the solution. For a more complete discussion of the problem we refer the reader to [**6**] and references therein.

In this paper we study the regularity of the solution to the principal-agent problem, for the special case when $b(x, y) = x \cdot y$ and $c(y) = |y|^2$, the C^1 regularity of the minimizer was proved by Carlier and Lachand-Robert [**5**], later Caffarelli and Lions [**2**] gave a very beautiful proof of $C^{1,1}$ regularity. For the general preference function, under some suitable condition we will adapt the method of Carlier and

2020 *Mathematics Subject Classification.* Primary 35J96, 35J25, 35B65.
Key words and phrases. Principal-agent problem, optimal transport.

Lachand-Robert to show that the minimizer is C^1. The proof is based on a perturbation argument, which is usually more difficult in the case of general $b(x, y)$. Since for the bilinear case $b(x, y) = x \cdot y$, the space of admissible functions is a subset of convex functions, then one can cut the graph of the minimizer by a hyperplane and replace the part below the hyperplane with the flat one. But for general $b(x, y)$ when one uses the natural choice of $b(\cdot, y) + \lambda$ (λ is a constant) instead of a hyperplane to perturb the minimizer, the shape of the domain where the function is perturbed is hard to control except for some special y.

2. Conditions, definitions and the main result

In this section, we will list the definitions and conditions which will be used later. The following (**B0**)–(**B3**) conditions were introduced into the principal-agent problem by Figalli, Kim and McCann [**6**].

Let $\overline{X}$ be the closure of a set $X \subset \mathbb{R}^n$. For each fixed $(x_0, y_0) \in \overline{X} \times \overline{Y}$ we assume:

(**B0**) $b \in C^4(\overline{X} \times \overline{Y})$, where $X \subset \mathbb{R}^n$ and $Y \subset \mathbb{R}^n$ are open and bounded;
(**B1**) (bi-twist) both $x \in \overline{X} \mapsto D_y b(x, y_0)$ and $y \in \overline{Y} \mapsto D_x b(x_0, y)$ are diffeomorphisms onto their ranges;
(**B2**) (bi-convexity) both $X_{y_0} := D_y b(X, y_0)$ and $Y_{x_0} := D_x b(x_0, Y)$ are convex subsets of $\mathbb{R}^n$.
(**B3**) (non-negative cross-curvature)

$$\frac{\partial^4}{\partial s^2 \partial t^2}\bigg|_{(s,t)=(0,0)} b(x(s), y(t)) \geq 0 \tag{2.1}$$

whenever either of the two curves $s \in [-1, 1] \mapsto D_y b(x(s), y(0))$ and $t \in [-1, 1] \mapsto D_x b(x(0), y(t))$ forms an affinely parameterized line segment ($\in X_{y(0)}$, or $\in Y_{x_0}$, respectively).

REMARK 2.1. The condition (**B3**) is a strengthened version of the MTW (or A3W) condition in the regularity theory of optimal transport with general cost function.

Now in order to formulate the principal-agent problem as a minimization problem over some space of admissible functions, we need the definition of b-convexity.

DEFINITION 1. A function $u : X \mapsto \mathbb{R}$ is called b-convex if $u = (u^{b^*})^b$, where

$$v^b(x) = \sup_{y \in \overline{Y}} \{b(x, y) - v(y)\}, \text{ and } u^{b^*}(y) = \sup_{x \in \overline{X}} \{b(x, y) - u(x)\}. \tag{2.2}$$

By the above definition and (**B0**), it is easy to see that a b-convex function is semi-convex, which implies that it is differentiable almost everywhere. In the following we will use $\mathrm{Dom}Du$ to denote the set where u is differentiable. Then by (**B1**), we can define the so called b-exponential map.

DEFINITION 2. For each $q \in Y_x$ we define $y_b(x, q)$ as the unique solution to

$$D_x b(x, y_b(x, q)) = q, \tag{2.3}$$

where the uniqueness is guaranteed by (**B1**).

Now suppose the price menu is given by the function $v(y)$, the agent x will try to choose the product that maximize the quantity $b(x, y) - v(y)$ among all $y \in \overline{Y}$. Then one can define a new function $u(x) = \sup_{y \in \overline{Y}} \{b(x, y) - v(y)\}$, which is b-convex by Definition 1. By (**B1**), we have $u(x) = b(x, y_b(x, Du(x))) - v(y_b(x, Du(x)))$ for all $x \in \mathrm{Dom}Du$. As discussed in section 4 of [**6**], the principal-agent problem is equivalent to the following minimization problem.

$$\min_{u \in \mathcal{U}_0} L(u), \tag{2.4}$$

where $\mathcal{U}_0 := \{u | u(x) \text{ is } b\text{-convex and } u \geq u_0(x) = b(x, y_0) - c(y_0)\}$, and

$$L(u) = \int_X [c(y_b(x, Du(x)) - b(x, y_b(x, Du(x))) + u] d\mu. \tag{2.5}$$

The point y_0 in the definition of $\mathcal{U}_0$ is the so called null product(or outside option), which the principal offers to all agents at zero profit. So $u_0(x)$ can be seen as the reservation utility of agent x, below which he will reject the principal's offer.

From now on, we will assume the density of different types of agents is given by $d\mu = f(x)dx$, where $f \in C^0(\overline{X}) \cap W^{1,\infty}(\overline{X})$ is a positive function. It was proved in [**6**] that if b satisfies (**B0**)–(**B3**), and if $c(y)$ is strictly b^*-convex, i.e., if $\mathrm{Dom}Dc^b = \overline{X}$, then there exists a unique solution to the above minimization problem. To state our regularity result, we need the following condition on $c(y)$, which is stronger than the strict b^*-convexity.

Condition 1. $c \in C^{1,1}(\overline{X})$, and $c(y_b(x,q)) - b(x, y_b(x,q))$ is strongly convex on the variable $q \in Y_x$ for $x \in \overline{X}$ uniformly, namely there exists a fixed $\delta > 0$ (independent of x), such that $c(y_b(x,q)) - b(x, y_b(x,q)) - \delta|q|^2$ is a convex function with respect to q.

Remark 2.2. By Proposition 4.4 in [**6**], if b satisfies (**B0**)–(**B3**), and if $c(y)$ is (strictly) b^*-convex, we have that $c(y_b(x,q)) - b(x, y_b(x,q))$ is a (strictly) convex function with respect to q.

Example 1. Given two functions F, G, and $\epsilon > 0$, let $b(x,y) = x \cdot y + \epsilon F(x)G(y)$. If $F \in C^4(\overline{X})$ and $G \in C^4(\overline{Y})$ are convex functions, then b satisfies (**B3**). Moreover, it satisfies (**B0**)–(**B1**) if $\sup_X |DF| < \frac{1}{\epsilon}$ and $\sup_Y |DG| < 1$. Lastly, (**B2**) is satisfied if the domains X, Y are sufficiently convex in the sense that all principal curvatures are sufficiently large at each boundary point. To generate examples satisfying Condition 1, let $c(y) = y^2$ and note that $y_b(x,q)$ satisfies $D_x b(x, y_b(x,q)) = q$; thus, $y_b(x,q) = q - \epsilon DF(x)G(y_b(x,q))$, and by letting $\psi(x,q) := c(y_b(x,q)) - b(x, y_b(x,q))$, we have

$$\begin{aligned}\psi(x,q) &= |q|^2 - 2\epsilon(q \cdot DF(x))G(y_b(x,q)) + \epsilon^2|DF(x)|^2 G(y_b(x,q)) - b(x, y_b(x,q)) \\ &= |q|^2 - x \cdot q + \epsilon\tilde{\psi}_\epsilon(x,q),\end{aligned}$$

where

$$\tilde{\psi}_\epsilon(x,q) = -2(q \cdot DF(x))G(y_b(x,q)) + \epsilon|DF(x)|^2 G(y_b(x,q)) + (x \cdot DF(x))G(y_b) - F(x)G(y_b).$$

The regularity imposed on F and G ensures the existence of $C_1 > 0$ and $C_2 > 0$ such that

$$|D^2\tilde{\psi}_\epsilon| \leq C_1 + \epsilon C_2.$$

Thus, if $\epsilon > 0$ satisfies

$$(\frac{1}{2} + C_1 + \epsilon C_2)\epsilon \leq 2$$

(e.g. $\epsilon = \min\{1, \frac{2}{\frac{1}{2} + C_1 + C_2}\}$), then we may take $\delta = \frac{\epsilon}{2}$ in Condition 1.

THEOREM 2.3. *If b satisfies* (**B0**)–(**B3**), *and if c satisfies Condition* 1, *then the unique solution to the above minimization problem is in $C^1(X)$.*

3. Proof of Theorem 1

The proof is based on a perturbation argument. Before embarking on the proof, we will give an interesting lemma that will be used later.

LEMMA 3.1 ([**5**]). *Suppose u is a convex function defined on a bounded convex domain $\Omega \subset \mathbb{R}^n$ which contains the origin 0, and $u(0) = 0$. If u is singular at 0 (namely $\partial u(0)$ contains more than one point) and $0 \in \mathrm{ri}\partial u(0)$, where $\mathrm{ri}\partial u(0)$ denotes the relative interior of the set $\partial u(0)$ (a set consists of all sub-gradients of u at 0). Then we have the estimate*

$$\int_{S_\epsilon} |Du(p)|^2 dp \geq C|S_\epsilon|, \tag{3.1}$$

where $S_\epsilon := \{p \in \Omega | u(p) \leq \epsilon\}$, $|S_\epsilon|$ is its volume, and C is a constant independent of ϵ.

This lemma was first proved by Carlier and Lachand-Robert in [**5**] by using a blow-up analysis and a compactness argument, because of the independent interest of this lemma, we will provide a more direct constructive proof of it, and in particular, the constant C can be explicitly estimated. The proof is very much inspired by [**2**].

PROOF OF LEMMA 1. Since $0 \in \mathrm{ri}\partial u(0)$, by rotating the coordinate system we can assume $u(p) \geq k|p_1|$ for some positive k, where $p = (p_1, p')$, and $p' = (p_2, p_3, \ldots, p_n)$. In the following, $\mathrm{Proj}(S_\epsilon)$ denotes the orthogonal projection of S_ϵ on the hyperplane $\{p | p_1 = 0\}$, and $\frac{1}{2}S_\epsilon$ denotes a $\frac{1}{2}$-dilation of S_ϵ with respect to the center 0. Now, we estimate the integration of $|Du|^2$ along the segment $I_{p'} := \{p | \mathrm{Proj}(p) = p', p \in S_\epsilon\}$, where $p' \in \mathrm{Proj}(\frac{1}{2}S_\epsilon)$. Let $\tilde{I}_{p'} := [a, a+d] \times \{p'\}$ be one of of the two components of $I_{p'} - \frac{1}{2}S_\epsilon$, without loss of generality we take the upper one, so by convexity we have $u(a, p') \leq \frac{1}{2}\epsilon$, and $u(a+d, p') = \epsilon$. Then we have

$$\begin{aligned}
\int_{I_{p'}} |Du|^2 dp_1 &\geq \int_{\tilde{I}_{p'}} |\frac{\partial u}{\partial p_1}|^2 dp_1 \\
&= \int_a^{a+d} |\frac{\partial u}{\partial p_1}|^2 dp_1 \\
&\geq (\frac{\frac{\epsilon}{2}}{d})^2 d = \frac{\epsilon^2}{4d},
\end{aligned}$$

the last "$\geq$" follows from the convexity of u. Since $u(p) \geq k|p_1|$, we have $S_\epsilon \subset [-\frac{\epsilon}{k}, \frac{\epsilon}{k}] \times \text{Proj}(S_\epsilon)$, and $d \leq \frac{2\epsilon}{k}$. Therefore

$$\begin{aligned}\int_{S_\epsilon} |Du|^2 dp &\geq \int_{\text{Proj}(\frac{1}{2}S_\epsilon)} \frac{\epsilon^2}{4\frac{2\epsilon}{k}} dp' \\ &\geq \frac{k}{8}\epsilon|\text{Proj}(\frac{1}{2}S_\epsilon)| \\ &= \frac{k}{2^{n+2}}\epsilon|\text{Proj}(S_\epsilon)| \\ &\geq \frac{k}{2^{n+2}}\epsilon\frac{|S_\epsilon|}{\frac{2\epsilon}{k}} = \frac{k^2}{2^{n+3}}|S_\epsilon|.\end{aligned}$$

So the Lemma 1 is proved with the constant $\frac{k^2}{2^{n+3}}$. □

Now we start the proof of Theorem 1, the main ingredient in the argument is that we do the change of variables as Figalli, Kim and McCann did in [**6**]. In the new variables, a b-convex function u in the original variables becomes a $\tilde{b}$-convex function with an extra property that it is also convex in the usual sense. Another advantage is that constant functions are $\tilde{b}$-convex, which allows us to perturb the minimizer by using constant functions safely.

Proof of Theorem 2.3. Firstly, we do a change of variable exactly same as that in [**6**]. By (**B0**)–(**B1**), the map $x \in \overline{X} \mapsto p = D_y b(x, y_0) \in \overline{X}_{y_0}$ and its inverse $p \in \overline{X}_{y_0} \mapsto x = x_b(y_0, p) \in \overline{X}$ are C^3 diffeomorphisms, where y_0 is the null product as in the minimization problem. Then $\tilde{u}(p) := u(x_b(y_0,p)) - b(x_b(y_0,p), y_0) + c(y_0)$ is a non-negative $\tilde{b}$-convex function, where $\tilde{b}(p, y) := b(x_b(y_0, p), y) - b(x_b(y_0, p), y_0) + c(y_0)$. In fact, the above correspondence between u and $\tilde{u}$ defines a $1-1$ map between the space $\mathcal{U}_0$ and the space $\widetilde{\mathcal{U}}_0$ of non-negative $\tilde{b}$-convex function on $\overline{X}_{y_0}$. By switching x and y in Remark 2.2, we see that $\tilde{u}(p)$ is a convex function. By a direct computation we see that

$$D_p\tilde{b}(p, y) = D_p x_b(y_0, p)\left(D_x b(x_b(y_0, p), y) - D_x b(x_b(y_0, p), y_0)\right), \tag{3.2}$$

and that

$$D_p\tilde{u} = D_p x_b(y_0, p)(Du(x_b(y_0, p)) - D_x b(x_b(y_0, p), y_0). \tag{3.3}$$

Hence, by (3.2) and (3.3) we have $y_{\tilde{b}}(p, D\tilde{u}(p)) = y_b\left(x_b(y_0, p), Du(x_b(y_0, p))\right)$. One should also notice that $\tilde{b}$ satisfies the same condition (**B0**)–(**B3**) as b does, except that $\tilde{b}$ might be only C^3 on the first variable.

Now, suppose $\tilde{u}$ is minimizer and it is not C^1 at a point $p_0 \in X_{y_0}$, namely $\partial\tilde{u}(p_0)$ contains more than one point. We claim that *one can find a point $y_1 \in \overline{Y}$, such that $D_p\tilde{b}(p_0, y_1) \in \text{ri}\partial\tilde{u}(p_0)$.*

Proof of the claim. Note that $\partial\tilde{u}(p_0)$ is a bounded convex set, so it is the convex hull of its extreme points. Since $\tilde{u}$ is a convex function, for any extreme point ω of $\partial\tilde{u}(p_0)$ we can find a sequence of points p_i at which u is differentiable, so that $p_i \to p_0$ and $D\tilde{u}(p_i) \to \omega$, as $i \to \infty$. By (**B1**), we have $D\tilde{u}(p_i) = D_p\tilde{b}(p_i, z_i)$ for a unique $z_i \in Y$. Therefore, by passing to a subsequence we can assume $z_i \to y_1$ for some $y_1 \in \overline{Y}$, then by taking limit we have $\omega = D_p(p_0, y_1)$. From the above discussion we see that all extreme points of $\partial\tilde{u}(p_0)$ are contained in $D_p\tilde{b}(p_0, \overline{Y})$, which is a convex set by (**B2**), so $\partial\tilde{u}(p_0) \subset D_p\tilde{b}(p_0, \overline{Y})$.

Then we perform the second change of variables. By (**B1**) we have that the C^2 diffeomorphism $p \in \overline{X}_{y_0} \mapsto \bar{p} := D_y\tilde{b}(p, y_1) \in \overline{X}_{y_1}$ and its inverse $\bar{p} \mapsto p_{\tilde{b}}(y_1, \bar{p})$ are well defined. Then $\bar{u}(\bar{p}) := \tilde{u}(p_{\tilde{b}}(y_1, \bar{p})) - \tilde{b}(p_{\tilde{b}}(y_1, \bar{p}), y_1) - \lambda_1$ is a $\bar{b}$-convex function, where $\lambda_1 = \tilde{u}(p_0) - \tilde{b}(p_0, y_1)$, and $\bar{b} := \tilde{b}(p_{\tilde{b}}(y_1, \bar{p}), y) - \tilde{b}(p_{\tilde{b}}(y_1, \bar{p}), y_1) - \lambda_1$ satisfies (**B0**)–(**B3**), except that it might be only C^2 on its first variable. By the same reason as in the first change of variables, $\bar{u}$ is not only $\bar{b}$-convex but also convex in the usual sense. Note that by the choice of y_1 in the above claim, $0 \in \mathrm{ri}\partial\bar{u}(\bar{p}_0)$, where $\bar{p}_0 = D_y\tilde{b}(p_0, y_1)$. In these new variables, the principal's net losses are given by

$$\widetilde{L}(\bar{u}) = \int_{X_{y_1}} [c(y_{\bar{b}}(\bar{p}, D\bar{u})) - \bar{b}(\bar{p}, y_{\bar{b}}(\bar{p}, D\bar{u})) + \bar{u}] f(x(\bar{p})) \det(\frac{\partial x^i}{\partial \bar{p}_j}) d\bar{p},$$

and the space of admissible functions becomes

$$\overline{\mathcal{U}}_0 = \{\bar{u} | \bar{u} \text{ is } \bar{b} - \text{convex and } \bar{u} \geq -\tilde{b}(p_{\tilde{b}}(y_1, \bar{p}), y_1) - \lambda_1\}.$$

Note that in the new variables, $c(y)$ still satisfies Condition 1, namely there exists some positive δ so that $c(y_{\bar{b}}(\bar{p}, q)) - \bar{b}(\bar{p}, y_{\bar{b}}(\bar{p}, q)) - \delta|q|^2$ is convex in the q variable for all $\bar{p} \in X_{y_1}$. It is also easy to see that after the above change of variables, constant functions are $\bar{b}$-convex.

In the following, for simplicity of notations, we will use $b(p, y)$ instead of $\bar{b}(\bar{p}, y)$, p instead of $\bar{p}$, u instead of $\bar{u}$, and $f(p)dp$ instead of $f(x(\bar{p})) \det(\frac{\partial x^i}{\partial \bar{p}_j}) d\bar{p}$. We will also omit the subscript $\bar{b}$. So the functional in the minimization problem looks like

$$\widetilde{L}(u) = \int_{X_{y_1}} [c(y(p, Du)) - b(p, y(p, Du)) + u] f(p) dp.$$

Now suppose u is a minimizer, by the second change of variables, we see that $u \geq 0$. Let $u_\epsilon := \max\{u, \epsilon\}$ and $S_\epsilon := \{u \leq \epsilon\}$. Since constant function ϵ is b-convex, u_ϵ is still in the space of admissible functions. Denote $G(p, q) = c(y(p, q)) - b(p, y(p, q)) - \delta|q|^2$. We have

$$\widetilde{L}(u_\epsilon) - \widetilde{L}(u) = I_1 + I_2 + I_3,$$

where

$$I_1 = -\delta \int_{S_\epsilon} |Du|^2 f(p) dp, \quad I_2 = \int_{S_\epsilon} (G(p, 0) - G(p, Du)) f(p) dp,$$

and

$$I_3 = \int_{S_\epsilon} (\epsilon - u) f(p) dp.$$

Since $0 \leq u \leq \epsilon$ in S_ϵ, we have $I_3 \leq C_1 \epsilon |S_\epsilon|$, for some constant C_1. For I_2, we have

$$\begin{aligned} I_2 &\leq \int_{S_\epsilon} D_q G(p, 0) \cdot D(\epsilon - u) f(p) dp \\ &= \int_{\partial S_\epsilon \cap \partial X_{y_1}} (\epsilon - u) f(p) D_q G(p, 0) \cdot \vec{n} \, d\xi \\ &\quad - \int_{S_\epsilon} (\epsilon - u) f(p) \operatorname{div}_p(D_q G(p, 0)) dp - \int_{S_\epsilon} (\epsilon - u) Df(p) \cdot D_q G(p, 0) dp, \end{aligned}$$

where the inequality follows from the convexity of G in q variable, and the equality follows from the divergence theorem. Since b is at least C^2 and f is $W^{1,\infty}$, all the integrand in the above three integrals are bounded by $C\epsilon$, for some constant C. For the area of $\partial S_\epsilon \cap \partial X_{y_1}$, we need to use a simple estimate

$$|\partial S_\epsilon \cap \partial X_{y_1}| \leq C|S_\epsilon|,$$

where C is a constant. This estimate was proved by Carlier and Lachand-Robert **[5]** and for reader's convenience we include their proof here.

PROOF OF THE ESTIMATE **[5]**.

$$|S_\epsilon| = \frac{1}{n}\int_{S_\epsilon} \operatorname{div}(p-p_0)dp \geq \frac{1}{n}\int_{\partial S_\epsilon \cap \partial X_{y_1}} (p-p_0)\cdot \vec{n}\, d\xi \geq \frac{1}{n}C|\partial S_\epsilon \cap \partial X_{y_1}|,$$

where p_0 is the assumed singular point of u, and the last inequality follows from the convexity of X_{y_1} (by the convexity, $(p-p_0)\cdot\vec{n}$ is bounded from below by some positive constant). By this estimate, we have $I_2 \leq C_2\epsilon|S_\epsilon|$, for some constant C_2. Then by Lemma 1, we have

$$\widetilde{L}(u_\epsilon) - \widetilde{L}(u) \leq -C_1|S_\epsilon| + C_2\epsilon|S_\epsilon| + C_3\epsilon|S_\epsilon|,$$

note that the constants C_1, C_2, C_3 are all independent of ϵ. When ϵ is sufficiently small, we see that $\widetilde{L}(u_\epsilon) < \widetilde{L}(u)$, which contradicts the fact that u is a minimizer. So u must be C^1. □

References

[1] M. Armstrong, *Multiproduct nonlinear pricing*, Econometrica **64** (1996) 51–75.

[2] Caffarelli and Lions, *Unpublished notes.*

[3] Guillaume Carlier, *A general existence result for the principal-agent problem with adverse selection*, J. Math. Econom. **35** (2001), no. 1, 129–150, DOI 10.1016/S0304-4068(00)00057-4. MR1817791

[4] J.-C. Rochet and P. Choné, *Ironing sweeping and multidimensional screening*, Econometrica **66** (1998) 783–826.

[5] G. Carlier and T. Lachand-Robert, *Regularity of solutions for some variational problems subject to a convexity constraint*, Comm. Pure Appl. Math. **54** (2001), no. 5, 583–594, DOI 10.1002/cpa.3. MR1811127

[6] Alessio Figalli, Young-Heon Kim, and Robert J. McCann, *When is multidimensional screening a convex program?*, J. Econom. Theory **146** (2011), no. 2, 454–478, DOI 10.1016/j.jet.2010.11.006. MR2888826

[7] Xi-Nan Ma, Neil S. Trudinger, and Xu-Jia Wang, *Regularity of potential functions of the optimal transportation problem*, Arch. Ration. Mech. Anal. **177** (2005), no. 2, 151–183, DOI 10.1007/s00205-005-0362-9. MR2188047

[8] R. Wilson, *Nonlinear Pricing*, Oxford University Press, Oxford, 1993.

School of Mathematical Sciences, University of Science and Technology of China, Hefei, 230026, People's Republic of China

Email address: `chenshib@ustc.edu.cn`

Contemporary Mathematics
Volume **781**, 2023
https://doi.org/10.1090/conm/781/15709

A doubly monotone flow for constant width bodies in $\mathbb{R}^3$

Ryan Hynd

ABSTRACT. We introduce a flow in the space of constant width bodies in three-dimensional Euclidean space that simultaneously increases the volume and decreases the circumradius of the shape as time increases. Starting from any initial constant width figure, we show that the flow exists for all positive times and converges to a closed ball as time tends to plus infinity. We also anticipate that this flow is interesting to study for negative times and that it would provide a mechanism to decrease the volume and increase the circumradius of a constant width body.

CONTENTS

1. Introduction

A constant width body is a compact, convex subset of Euclidean space in which parallel supporting planes are separated by the same distance in every direction. For such a shape, this distance is called its width. In this note, we will focus exclusively on constant width bodies whose respective widths are equal to one and simply refer to them as having constant width. The simplest example of a constant width body is a (closed) ball of radius 1/2. It is also known that balls of radius 1/2 enclose the most volume among constant width shapes.

It is natural to inquire about volume-minimizing constant width bodies. In the plane, it was proved independently by Lebesgue [**33, 34**] and Blaschke [**10, 11**] over a century ago that Reuleaux triangles enclose the least amount of area. A Reuleaux triangle is the intersection of three closed disks of radius 1 which are centered at

2020 *Mathematics Subject Classification.* Primary 47J35, 52A38, 52A40.
This research was partially supported by NSF award DMS-1554130.

the vertices of an equilateral triangle of side length one. There have been many subsequent proofs of the Lebesgue-Blasckhe theorem including [**15**, **20**, **23**–**25**, **36**]. We also note that Harrell [**25**] showed that Reuleaux triangles are uniquely area-minimizing among constant width shapes.

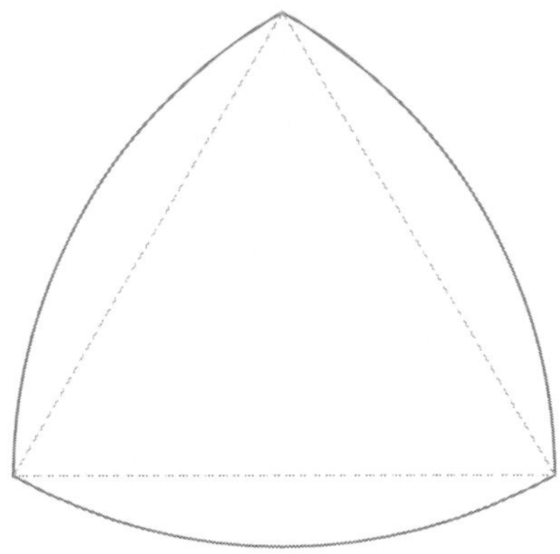

FIGURE 1. The boundary curve of a Reuleaux triangle with inscribed equilateral triangle.

For constant width bodies in dimensions larger than two, it is known that volume-minimizing constant width bodies exist. This can be seen as a consequence of the Blaschke selection theorem. While there have been some notable work on this topic such as [**2**, **3**, **8**, **14**, **15**, **50**], surprisingly little is known about these shapes. However, there are conjectured volume-minimizing constant width bodies in three-dimensional Euclidean space [**13**, **19**, **28**, **30**, **37**]. These are the bodies that Meissner (and Schilling) constructed [**38**] which are based on a regular tetrahedron, somewhat analogous to how the Reuleaux triangle is based on an equilateral triangle.

First Meissner considered a Reuleaux tetrahedron, which is the intersection of four balls of radius one each centered at the vertices of a regular tetrahedron. This figure has four faces, four vertices and six edges just like the regular tetrahedron. After realizing the Reuleaux tetrahedron does not have constant width, Meissner was able to round three of the six edges in two ways to obtain two distinct constant width bodies; we recommend diagrams 106 and 107 of [**28**] for a detailed description of these procedures. Both of Meissner's tetrahedra have the same volume and surface area, and it has been long thought that these are volume-minimizing shapes. There have also been at least two numerical studies [**4**, **45**] which support this conjecture.

An interesting feature of a constant width body is that its inball and circumball are concentric. Moreover, the radii of these two balls sum to one. It turns out that any body of constant width which includes the regular simplex of diameter one necessarily has the largest possible circumradius (and therefore the smallest possible inradius). For example, the Reuleaux triangle has this property and so do Meissner's tetrahedra. It then seems reasonable to investigate the following question.

QUESTION. *Is there a connection between least volume bodies of constant width and those having largest circumradius?*

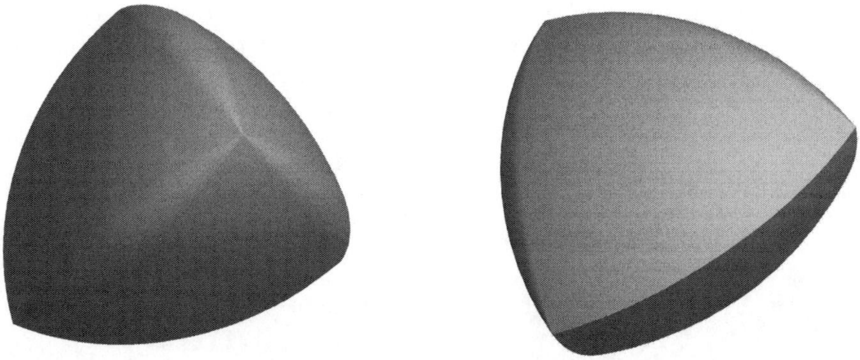

FIGURE 2. A Meissner tetrahedron in which three rounded edges meet in a vertex.

In an attempt to develop an approach to this question, we will propose a flow in the space of constant width bodies in $\mathbb{R}^3$ which has two distinctive features: when time moves forward, the volume increases and the circumradius decreases along the flow. We expect that as time tends to infinity, the flow would deform any starting shape into a ball of radius 1/2. We will investigate the existence of this flow and its behavior for large times in detail below. Nevertheless, we present this flow as a possible device which can be used to answer our motivating question. Namely, upon reversing time, we wonder if limiting shapes exist and lead us to some insight on understanding a possible relationship between least volume bodies of constant width and those having largest circumradius.

1.1. Quantities of interest. In what follows, we will employ the support function

$$H(u) = \max_{x \in K} x \cdot u \quad (u \in \mathbb{R}^3)$$

of a constant width body $K \subset \mathbb{R}^3$. As H is positively homogeneous, it is determined by its restriction to the unit sphere $h = H|_{\mathbb{S}^2}$. The constant width property of K is equivalent to h satisfying

$$h(u) + h(-u) = 1$$

for all $u \in \mathbb{S}^2$. It will also be convenient to consider

$$g = h - 1/2,$$

which is an odd function on $\mathbb{S}^2$.

Later in this note, we will show that the circumradius of K can be expressed in terms of g as

$$R(K) = \frac{1}{2} + \min_{a \in \mathbb{R}^3} \max_{|u|=1} |g(u) + a \cdot u|.$$

Likewise, we will explain that volume enclosed by K can be written in terms of g as

$$V(K) = \frac{\pi}{6} - \frac{1}{2} \int_{\mathbb{S}^2} \left(\frac{1}{2} |\nabla g|^2 - g^2 \right) d\sigma.$$

Here σ is two-dimensional Hausdorff measure on $\mathbb{R}^3$ normalized so that $\sigma(\mathbb{S}^2) = 4\pi$.

In addition, we will recall that the Hausdorff distance $d_{\mathcal{H}}(K_1, K_2)$ between two convex bodies $K_1, K_2 \subset \mathbb{R}^3$ is given by the supremum norm over $\mathbb{S}^2$ of the difference of their respective support functions. It follows that

$$d_{\mathcal{H}}(K_1, K_2) = \max_{|u|=1} |g_1(u) - g_2(u)|.$$

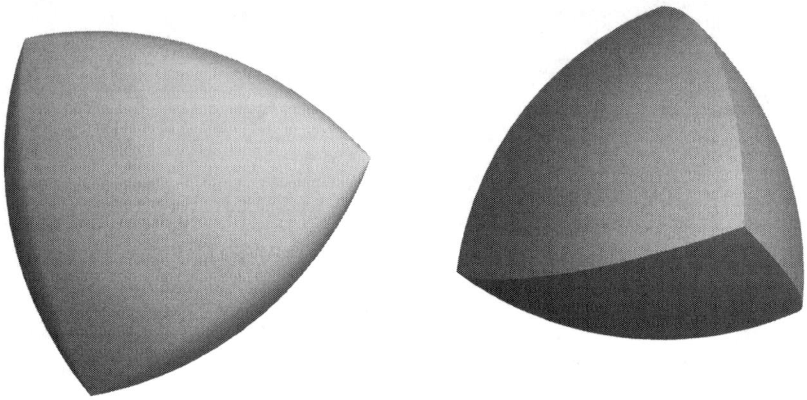

FIGURE 3. A Meissner tetrahedron having three rounded edges that form a triangle.

1.2. Notation and preliminaries. We will use the term "ball" to mean a closed ball and write

$$B_r(a) = \{x \in \mathbb{R}^3 : |x - a| \leq r\}.$$

The space $C(\mathbb{S}^2)/\mathcal{P}$. We will denote $C(\mathbb{S}^2)$ as the space of continuous $g : \mathbb{S}^2 \to \mathbb{R}$ endowed with the supremum norm

$$\|g\| = \max_{|u|=1} |g(u)|.$$

Let us also write $\mathcal{P} \subset C(\mathbb{S}^2)$ for the closed subspace of functions

$$\mathbb{S}^2 \to \mathbb{R}; u \mapsto a \cdot u$$

where a ranges over points in $\mathbb{R}^3$. This corresponds to convex bodies which are singletons. Since the volume and circumradius are invariant under translations, we will consider the quotient space

$$C(\mathbb{S}^2)/\mathcal{P} = \{g + \mathcal{P} : g \in C(\mathbb{S}^2)\},$$

which is endowed with the quotient norm

$$\|g + \mathcal{P}\| = \inf_{a \in \mathbb{R}^3} \max_{|u|=1} |g(u) + a \cdot u|.$$

For ease of notation, we will use the variable $\mathbf{g}$ to denote an element $C(\mathbb{S}^2)/\mathcal{P}$ whenever there is $g \in \mathbf{g}$ so that

$$\mathbf{g} = g + \mathcal{P}.$$

The space $\mathcal{P}^\perp$. We recall that the continuous dual space of $C(\mathbb{S}^2)/\mathcal{P}$ when endowed with operator norm is isometrically isomorphic to

$$\mathcal{P}^\perp = \{\xi \in M(\mathbb{S}^2) : \xi|_{\mathcal{P}} = 0\}.$$

Here $M(\mathbb{S}^2) = C(\mathbb{S}^2)^*$ is the space of signed Radon measures on $\mathbb{S}^2$ endowed with the total variation norm. In particular, $\mathcal{P}^\perp$ admits the norm

$$\|\xi\|_* = \sup\left\{\langle \xi, g\rangle : \mathbf{g} \in C(\mathbb{S}^2)/\mathcal{P},\ \|\mathbf{g}\| \le 1\right\},$$

where

$$\langle \xi, g\rangle := \int_{\mathbb{S}^2} g d\xi$$

is the natural pairing between $\xi \in M(\mathbb{S}^2)$ and $g \in C(\mathbb{S}^2)$.

We will denote $\mathcal{J}$ as the subdifferential of

$$C(\mathbb{S}^2)/\mathcal{P} \ni \mathbf{g} \mapsto \frac{1}{2}\|\mathbf{g}\|^2$$

and $\mathcal{J}^*$ for the subdifferential of

$$\mathcal{P}^\perp \ni \xi \mapsto \frac{1}{2}\|\xi\|_*^2.$$

By convex duality, $\xi \in \mathcal{J}(\mathbf{g})$ if and only if $\mathbf{g} \in \mathcal{J}^*(\xi)$ if and only if

$$\|\mathbf{g}\|^2 = \langle \xi, g\rangle = \|\xi\|_*^2. \tag{1.1}$$

The space $\mathcal{C}$. Another natural space for us to study is

$$\mathcal{C} := \left\{g \in C(\mathbb{S}^2) : g + \frac{1}{2} = H|_{\mathbb{S}^2}, H \text{ is the support function of a constant width body in } \mathbb{R}^3\right\}. \tag{1.2}$$

We will identify a useful compactness property of $\mathcal{C}$ below. And as mentioned above, if K is a constant width body associated with $g \in \mathcal{C}$, then its circumradius is given by

$$R(K) = \frac{1}{2} + \|\mathbf{g}\|.$$

In addition, we will consider the functional defined as

$$E(\mathbf{g}) := \begin{cases} \int_{\mathbb{S}^2} \left(\frac{1}{2}|\nabla g|^2 - g^2\right) d\sigma, & g \in \mathcal{C} \\ +\infty, & g \notin \mathcal{C} \end{cases}$$

for $\mathbf{g} \in C(\mathbb{S}^2)/\mathcal{P}$. We note that E is well-defined since the integral $\int_{\mathbb{S}^2}\left(\frac{1}{2}|\nabla g|^2 - g^2\right) d\sigma$ is invariant under the translation of g by elements in $\mathcal{P}$. Moreover, it is not hard to see that E is convex, proper, and lower-semicontinuous. And as we previously noted,

$$V(K) = \frac{\pi}{6} - \frac{1}{2}E(\mathbf{g})$$

for a constant width body K associated with $\mathbf{g}$. Since $V(K) \ge 0$, we also have

$$\sup_{\mathbf{g} \in \mathcal{C}} E(\mathbf{g}) \le \frac{\pi}{3}. \tag{1.3}$$

Notice that if $g_1, g_2 \in \mathcal{C}$ with corresponding constant with bodies $K_1, K_2 \subset \mathbb{R}^3$, then

$$\|\mathbf{g}_1 - \mathbf{g}_2\| = \inf_{a \in \mathbb{R}^3} d_{\mathcal{H}}(K_1 + a, K_2).$$

Here $d_{\mathcal{H}}$ is the Hausdorff distance. In particular, if $\mathbf{g}_n \to \mathbf{g}$ and if K_n and K are respective convex bodies associated with $g_n \in \mathcal{C}$ and $g \in \mathcal{C}$, there is a sequence $(a_n)_{n\in\mathbb{N}} \subset \mathbb{R}^3$ for which

$$\lim_{n\to\infty} \|\mathbf{g}_n - \mathbf{g}\| = \lim_{n\to\infty} d_{\mathcal{H}}(K_n + a_n, K) = 0.$$

That is, up to translations, K_n converges to K in the Hausdorff topology.

1.3. A doubly nonlinear evolution. In what follows, we will study solutions $\xi : [0,\infty) \to \mathcal{P}^{\perp}$ of the *doubly nonlinear evolution*

$$\partial E^*(\dot{\xi}(t)) + \mathcal{J}^*(\xi(t)) \ni 0 \quad \text{a.e. } t \geq 0 \tag{1.4}$$

for a given initial condition. Here

$$E^*(\zeta) := \sup\{\langle \zeta, g\rangle - E(\mathbf{g}) : g \in \mathcal{C}\}, \quad (\zeta \in \mathcal{P}^{\perp})$$

is the convex dual of E and we recall $\mathcal{J}^*$ is the subdifferential of $\frac{1}{2}\|\cdot\|_*^2$. It is typical to consider a solution of (1.4) to mean that ξ is absolutely continuous and that there is a measurable $\mathbf{g} : [0,\infty) \to C(\mathbb{S}^2)/\mathcal{P}$ with

$$\mathbf{g}(t) \in \mathcal{J}^*(\xi(t)) \cap (-\partial E^*(\dot{\xi}(t))) \tag{1.5}$$

for almost every $t \geq 0$. Let us for the moment suppose we have such a solution.

First monotonicity formula. By direct computation, we find

$$\frac{d}{dt}\frac{1}{2}\|\xi(t)\|_*^2 = -\left[E^*(\dot{\xi}(t)) + E(\mathbf{g}(t))\right]$$

for almost every $t \geq 0$. Integrating this formula on the interval $[s,t]$ gives

$$\frac{1}{2}\|\xi(s)\|_*^2 = \frac{1}{2}\|\xi(t)\|_*^2 + \int_s^t E^*(\dot{\xi}(\tau))d\tau + \int_s^t E(\mathbf{g}(\tau))d\tau.$$

This identity, combined with (1.1) and (1.5), implies

$$\|\mathbf{g}(t)\| = \|\xi(t)\|_* \text{ is a nonincreasing function of t.}$$

Second monotonicity formula. In view of (1.5), we also have $-\dot{\xi}(t) \in \partial E(\mathbf{g}(t))$ for almost every $t \geq 0$. If, in addition $g : [0,\infty) \to C(\mathbb{S}^2); t \mapsto g(t)$ is differentiable almost everywhere, then

$$\frac{d}{dt}E(\mathbf{g}(t)) = -\langle \dot{\xi}(t), \dot{g}(t)\rangle$$

for almost every $t \geq 0$. Using (1.1), we find

$$\begin{aligned}
&\langle \xi(t+\tau) - \xi(t), g(t+\tau) - g(t)\rangle \\
&= \|\xi(t+\tau)\|_*^2 - \langle \xi(t), g(t+\tau)\rangle - \langle \xi(t+\tau), g(t)\rangle + \|\xi(t)\|_*^2 \\
&\geq \|\xi(t+\tau)\|_*^2 - \|\xi(t)\|_*\|\mathbf{g}(t+\tau)\| - \|\xi(t+\tau)\|_*\|\mathbf{g}(t)\| + \|\xi(t)\|_*^2 \\
&= \|\xi(t+\tau)\|_*^2 - 2\|\xi(t)\|_*\|\xi(t+\tau)\|_* + \|\xi(t)\|_*^2 \\
&= \left(\|\xi(t+\tau)\|_* - \|\xi(t)\|_*\right)^2.
\end{aligned}$$

Thus, we expect

$$\frac{d}{dt}E(\mathbf{g}(t)) = -\langle \dot{\xi}(t), \dot{g}(t)\rangle \leq -\left(\frac{d}{dt}\|\xi(t)\|_*\right)^2$$

for almost every $t \geq 0$. In particular,

$$E(\mathbf{g}(t)) \text{ is a nonincreasing function of } t.$$

Large time limits. Since $E(\mathbf{g}(t))$ is nonnegative, nonincreasing, and integrable on $[0, \infty)$, it must be that

$$\lim_{t\to\infty} tE(\mathbf{g}(t)) = 0.$$

Using this limit and the compactness of $\mathcal{C}$, we would then be able to conclude

$$\lim_{t\to\infty} \|\mathbf{g}(t)\| = 0.$$

Geometric interpretation. For each $t \geq 0$, $g(t) + 1/2$ is the $\mathbb{S}^2$ restriction of the support function of a constant width body $K_t \subset \mathbb{R}^3$. Given the monotonicity formulae above and the way circumradius and volume can be expressed in terms of the support function,

$$R(K_t) \text{ is a nonincreasing function of } t$$

and

$$V(K_t) \text{ is a nondecreasing function of } t.$$

In view of the above large time limits,

$$\begin{cases} \lim_{t\to\infty} t\left(\frac{\pi}{6} - V(K_t)\right) = 0 \\ \\ \lim_{t\to\infty} R(K_t) = \frac{1}{2}. \end{cases}$$

And up to translations,

$$K_t \text{ converges to } B_{1/2}(0) \text{ in the Hausdorff topology.}$$

Approach to existence. Our goal is to establish that, for given initial conditions $g^0 \in \mathcal{C}$ and $\xi^0 \in \mathcal{P}^\perp$ such that

$$\xi^0 \in \mathcal{J}(\mathbf{g}^0),$$

there is a solution of (1.4) $\xi : [0, \infty) \to \mathcal{P}^\perp$ as described above. Then we could obtain a mapping $\mathbf{g} : [0, \infty) \to C(\mathbb{S}^2)/\mathcal{P}$ and make the analogous geometric conclusions. However, we do not know how to carry this out primarily because the functional E^* is not coercive. In particular, having an a priori bound on the integral

$$\int_0^\infty E^*(\dot{\xi}(t))dt \tag{1.6}$$

does not suggest that we can construct an absolutely continuous solution ξ.

This is a typical problem encountered in the study of "rate-independent" doubly nonlinear evolutions [**18, 31, 35, 40–44, 48**]. For these flows, E^* is usually a norm, so the gradient or subdifferential of E^* is homogeneous of degree 0. It turns out that solutions to these flows have bounded variation. While our functional E^* is not a norm, it is convex, lower-semicontinuous and proper. Using this basic information, we will develop the notion of the E^* variation of a mapping $\xi : [0, \infty) \to \mathcal{P}^\perp$. Replacing E^* with a norm in this definition results in the usual variation of ξ; and if ξ is absolutely continuous, the E^* variation of ξ on $[0, \infty)$ is equal to the integral (1.6).

These considerations lead to a notion of weak solution based on the "classical" notion of solution described above. We will show that classical solutions are weak solutions. Moreover, we will prove that weak solutions ξ exist for given initial conditions and their companion mappings $\mathbf{g}$ essentially satisfy the properties discussed above. To this end, we will use a compactness based approach by designing an approximation sequence, establishing various bounds on these approximations, and then extracting a subsequence which converges to a solution with desired properties. In order to carry out this procedure, we will need to first develop some ideas for support functions and to discuss the appropriate function spaces. These are the topics of the subsequent sections of this paper.

2. Support functions

Let us recall a few basic facts about convex bodies. Suppose $K \subset \mathbb{R}^3$ is a convex body with support function

$$H(u) = \max_{x \in K} x \cdot u \quad (u \in \mathbb{R}^3).$$

Observe that H is positively homogeneous and convex. Moreover,

$$K = \bigcap_{|u|=1} \left\{ x \in \mathbb{R}^3 : x \cdot u \le H(u) \right\}.$$

This identity can be used to show that the class of positively homogeneous, convex functions $H : \mathbb{R}^3 \to \mathbb{R}$ are in one-to-one correspondence with convex bodies.

If K_1 and K_2 are two convex bodies with respective support functions H_1 and H_2, we may consider their Minkowski sum

$$K_1 + K_2 := \{x_1 + x_2 \in \mathbb{R}^3 : x_1 \in K_1, x_2 \in K_2\},$$

which is also convex body with support function $H_1 + H_2$. In addition, we note that $K_1 \subset K_2$ if and only if $H_1 \le H_2$.

The Hausdorff distance

$$d_{\mathcal{H}}(K_1, K_1) := \inf\{r > 0 : K_1 \subset K_2 + B_r(0),\ K_2 \subset K_1 + B_r(0)\}$$

between K_1 and K_2 may also be expressed in terms of their support functions. Indeed

$$\begin{aligned} d_{\mathcal{H}}(K_1, K_1) &= \inf\{r>0 : H_{K_1}(u) \le H_{K_2}(u) + r \text{ and } H_{K_1}(u) \le H_{K_2}(u) + r \text{ for } |u|=1\} \\ &= \inf\left\{ r > 0 : \max_{|u|=1} |H_{K_1}(u) - H_{K_2}(u)| \le r \right\} \\ &= \max_{|u|=1} |H_{K_1}(u) - H_{K_2}(u)|. \end{aligned}$$

Many more properties of support functions can be found in the standard references on convex bodies such as [**13, 47, 49**].

In this section, we will study a fixed constant width body K with support function H. We note that the equality

$$K + (-K) = B_1(0), \tag{2.1}$$

is equivalent to K having constant width. This in turn holds if and only if

$$H(u) + H(-u) = |u| \tag{2.2}$$

for each $u \in \mathbb{R}^3$. Below, we will derive a formula for the circumradius and volume of K in terms of H. We will also derive a few estimates on H which will imply a compactness property for constant width bodies.

We note that most if not all of the following results are likely to be found in the literature on constant width bodies (including [**13, 28, 37**]). However, we have included them in attempt to present a somewhat unified treatment of the support functions of constant width bodies. Moreover, many of the results are valid in arbitrary dimension. Nevertheless, we will only focus on constant width bodies in $\mathbb{R}^3$.

2.1. In and circumradius. We define the *inradius* of K to be

$$r(K) = \sup\{r \ge 0 : K \supset B_r(a) \text{ some } a \in K\},$$

and an *inball* as any ball with $B_{r(K)}(a) \subset K$. Namely, an inball is a ball of maximal radius which can be included in K.

PROPOSITION 2.1.

$$r(K) = \max_{a\in\mathbb{R}^3}\min_{|u|=1}\{H(u) - a\cdot u\},$$

and the maximum occurs at some $a \in K$. Moreover, $B_{r(K)}(a)$ is an inball.

PROOF. Note that $u \mapsto a\cdot u + r|u|$ is the support function of $B_r(a)$. Therefore,

$$\begin{aligned} r(K) &= \sup\{r \ge 0 : K \supset B_r(a) \text{ some } a \in K\}\\ &= \sup\{r \ge 0 : a\cdot u + r \le H(u) \text{ all } |u| = 1 \text{ and some } a \in K\}\\ &= \sup\left\{r \ge 0 : r \le \min_{|u|=1}\{H(u) - a\cdot u\} \text{ for some } a \in K\right\}\\ &= \sup\left\{r \ge 0 : r \le \max_{a\in K}\min_{|u|=1}\{H(u) - a\cdot u\}\right\}\\ &= \max_{a\in K}\min_{|u|=1}\{H(u) - a\cdot u\}. \end{aligned}$$

If we choose $a \in K$ so that

$$r(K) = \min_{|u|=1}\{H(u) - a\cdot u\},$$

then $r(K) + a\cdot u \le H(u)$ for all $|u| = 1$. That is, $B_{r(K)}(a) \subset K$.

We next claim that, $r(K)$ is equal to

$$\overline{r}(K) := \sup_{a\in\mathbb{R}^3}\min_{|u|=1}\{H(u) - a\cdot u\}.$$

So far we have $r(K) \le \overline{r}(K)$. Moreover, for any $a \in \mathbb{R}^3$ there is a unit vector u for which $a\cdot u \ge 0$. Thus

$$\min_{|u|=1}\{H(u) - a\cdot u\} \le \max_{|u|=1} H(u) < \infty.$$

It follows that $\overline{r}(K) < \infty$. We may suppose $\overline{r}(K) > 0$ or else $\overline{r}(K) = r(K) = 0$. In this case, let $\epsilon \in (0, \overline{r}(K))$ and choose a^ϵ such that

$$\overline{r}(K) - \epsilon \le \min_{|u|=1}\{H(u) - a^\epsilon\cdot u\}.$$

Then $B_{\overline{r}(K)-\epsilon}(a^\epsilon) \le K$. In particular, $a^\epsilon \in K$ and

$$\overline{r}(K) - \epsilon \le r(K).$$

□

Analogously, we can define the *circumradius* of K as

$$R(K) = \inf\{r > 0 : K \subset B_r(a) \text{ some } a \in \mathbb{R}^3\}.$$

Note that since K is compact, $R(K) < \infty$. We'll also call a *circumball* a ball such that $K \subset B_{R(K)}(a)$. First we show that circumballs are unique.

LEMMA 2.2. *There can be only one $a \in \mathbb{R}^3$ such that $K \subset B_{R(K)}(a)$.*

PROOF. Set $r := r(K)$ and suppose $K \subset B_r(a_1) \cap B_r(a_2)$ with $a_1 \neq a_2$. For any $z \in K$, $|z - a_1| \leq r$ and $|z - a_2| \leq r$. Therefore,

$$|a_1 - a_2| < |a_1 - z| + |z - a_2| \leq 2r.$$

Set

$$s := \sqrt{r^2 - \left|\frac{a_1 - a_2}{2}\right|^2},$$

and note that $0 < s < r$. Moreover, for $z \in K$

$$\left|z - \frac{a_1 + a_2}{2}\right|^2 = \frac{1}{2}|z - a_1|^2 + \frac{1}{2}|z - a_2|^2 - \left|\frac{a_1 - a_2}{2}\right|^2 \leq r^2 - \left|\frac{a_1 - a_2}{2}\right|^2 = s^2.$$

Thus, $K \subset B_s((a_1 + a_2)/2)$. This contradicts the assumption that $r = r(K)$. Thus, it must be that $a_1 = a_2$. $\square$

In analogy with our formula for the inradius, we have the following formula for the circumradius.

PROPOSITION 2.3.

$$R(K) = \min_{a \in \mathbb{R}^3} \max_{|u|=1} \{H(u) - a \cdot u\}. \tag{2.3}$$

Moreover, $B_{R(K)}(a)$ is the circumball provided that

$$R(K) = \max_{|u|=1} \{H(u) - a \cdot u\}.$$

PROOF. Observe that if $K \subset B_r(a)$ for some $a \in \mathbb{R}^3$, then $H(u) \leq a \cdot u + r$ for all $|u| = 1$. As a result,

$$\begin{aligned} r &\geq \max_{|u|=1} \{H(u) - a \cdot u\} \\ &\geq \inf_{a \in \mathbb{R}^3} \max_{|u|=1} \{H(u) - a \cdot u\} \\ &=: \overline{R}(K). \end{aligned}$$

Thus $R(K) \geq \overline{R}(K)$. Now let $\epsilon > 0$ and select $a^\epsilon \in \mathbb{R}^3$ for which

$$\overline{R}(K) + \epsilon \geq \max_{|u|=1} \{H(u) - a^\epsilon \cdot u\}.$$

This implies, $K \subset B_{\overline{R}(K)+\epsilon}(a^\epsilon)$. Thus,

$$R(K) \leq \overline{R}(K) + \epsilon.$$

It follows that $R(K) = \overline{R}(K)$.

Suppose $(a^k)_{k \in \mathbb{N}}$ is a sequence in $\mathbb{R}^3$ for which

$$R(K) = \lim_{k \to \infty} \max_{|u|=1} \{H(u) - a^k \cdot u\}.$$

Then for all $|v| = 1$ and sufficiently large $k \in \mathbb{N}$,

$$\begin{aligned} -a^k \cdot v &= H(v) - a^k \cdot v - H(v) \\ &\leq \max_{|u|=1}\{H(u) - a^k \cdot u\} + \max_{|u|=1} |H(u)| \\ &\leq R(K) + 1 + \max_{|u|=1} |H(u)|. \end{aligned}$$

This implies that $(a^k)_{k\in\mathbb{N}}$ is bounded and therefore has a convergent subsequence $(a^{k_j})_{j\in\mathbb{N}}$ with limit a^∞. As a result,

$$R(K) = \lim_{j\to\infty} \max_{|u|=1}\{H(u) - a^{k_j} \cdot u\} = \max_{|u|=1}\{H(u) - a^\infty \cdot u\}.$$

We conclude (2.3). Since $H(u) \leq R(K) + a^\infty \cdot u$ it must also be that $K \subset B_{R(K)}(a^\infty)$. □

We will now make use of (2.2).

LEMMA 2.4. *K has a unique inball $B_{r(K)}(a)$ which is concentric with its circumball. Moreover,*

$$r(K) + R(K) = 1.$$

PROOF. Since $H(u) + H(-u) = 1$ for each $|u| = 1$,

$$\begin{aligned} r(K) &= \max_{a\in\mathbb{R}^3} \min_{|u|=1} \{H(u) - a \cdot u\} \\ &= \max_{a\in\mathbb{R}^3} \min_{|u|=1} \{1 - H(-u) - a \cdot u\} \\ &= 1 + \max_{a\in\mathbb{R}^3} \min_{|u|=1} \{-H(-u) - a \cdot u\} \\ &= 1 - \min_{a\in\mathbb{R}^3} \max_{|u|=1}\{H(-u) + a \cdot u\} \\ &= 1 - \min_{a\in\mathbb{R}^3} \max_{|u|=1}\{H(u) - a \cdot u\} \\ &= 1 - R(K). \end{aligned}$$

Now select $a \in K$ for which

$$r(K) = \min_{|u|=1} \{H(u) - a \cdot u\}.$$

This implies $H(u) \geq a \cdot u + r(K)$ for $|u| = 1$ so that $B_{r(K)}(a) \subset K$. In addition, we have

$$\begin{aligned} R(K) &= 1 - r(K) \\ &= 1 - \min_{|u|=1} \{H(u) - a \cdot u\} \\ &= \max_{|u|=1}\{1 - H(u) + a \cdot u\} \\ &= \max_{|u|=1}\{H(-u) + a \cdot u\} \\ &= \max_{|u|=1}\{H(u) - a \cdot u\}. \end{aligned}$$

Thus, $R(K) + a \cdot u \geq H(u)$ for all $|u| = 1$. Furthermore, $K \subset B_{R(K)}(a)$ and a is uniquely specified. □

Corollary 2.5.

$$R(K) \le \sqrt{\frac{3}{8}} \quad \textit{and} \quad r(K) \ge 1 - \sqrt{\frac{3}{8}}.$$

Equality holds in either inequality if K is a Meissner tetrahedron.

Proof. In view of (2.1), K has diameter 1. Jung's theorem [**29, 46**] implies

$$R(K) \le \sqrt{\frac{3}{2(3+1)}} = \sqrt{\frac{3}{8}}$$

and that equality would hold if K is a regular tetrahedron. Now let $M \subset \mathbb{R}^3$ be a Meissner tetrahedron. Then M includes a regular tetrahedron T of diameter 1. Therefore,

$$R(M) \ge R(T) = \sqrt{\frac{3}{8}}.$$

By the previous lemma, we also have

$$r(K) = 1 - R(K) \ge 1 - \sqrt{\frac{3}{8}}$$

and similarly conclude that equality holds if K is a Meissner tetrahedron. □

We can now derive a key formula for the circumradius of a body of constant width.

Proposition 2.6.

$$R(K) = 1/2 + \min_{a\in\mathbb{R}^3} \max_{|u|=1} |H(u) - 1/2 - a\cdot u|$$

and the minimum occurs at the center of the circumball for K.

Proof. Let $a \in K$ be the center of K's circumball. Then

$$a\cdot u + r(K) \le H(u) \le a\cdot u + R(K)$$

for all $|u| = 1$. As $r(K) + R(K) = 1$,

$$|H(u) - 1/2 - a\cdot u| \le R(K) - 1/2.$$

Thus,

$$\max_{|u|=1} |H(u) - 1/2 - a\cdot u| \le R(K) - 1/2.$$

Also note

$$\begin{aligned} R(K) - 1/2 &= \inf_{b\in\mathbb{R}^3} \max_{|u|=1} \{H(u) - 1/2 - b\cdot u\} \\ &\le \inf_{b\in\mathbb{R}^3} \max_{|u|=1} |H(u) - 1/2 - b\cdot u|. \end{aligned}$$

□

Remark 2.7. A corollary of Proposition 2.6 is

$$R(K) = \frac{1}{2} + d_{\mathcal{H}}(K - a, B_{1/2}(0)), \tag{2.4}$$

where a is the center of K's circumball.

Corollary 2.8. *If $B_{R(K)}(a)$ is the circumball of K, then*

(i) $|H(u) - 1/2 - a\cdot u| \le \sqrt{\frac{3}{8}} - \frac{1}{2}$ *for* $|u| = 1$, *and*

(ii) $|H(u) - H(v) - a\cdot(u-v)| \le |u-v|$ *for* $u, v \in \mathbb{R}^3$.

PROOF. *(i)* This follows from Corollary 2.5 and Proposition 2.6. *(ii)* Choose $x \in K$ for which $H(u) = x\cdot u$. As the diameter of K is 1,

$$\begin{aligned} H(u) - H(v) - a\cdot(u-v) &\le x\cdot u - x\cdot v - a\cdot(u-v) \\ &= (x-a)\cdot(u-v) \\ &\le |u-v|. \end{aligned}$$

The same argument implies $H(v) - H(u) - a\cdot(v-u) \le |u-v|$. □

2.2. A gradient estimate. Since H is convex, it is twice differentiable almost everywhere by Rademacher's theorem. In view of (2.2), we additionally have

$$0 \le D^2H(u)w\cdot w \le \frac{1}{|u|}$$

for almost every $u \in \mathbb{R}^3\setminus\{0\}$ and each $w \in \mathbb{S}^2$. This in turn implies that H is continuously differentiable away from 0 [**27**]. We can also quantify this below with the following estimate.

PROPOSITION 2.9. *For* $|u|, |v| \ge r$,

$$|DH(u) - DH(v)| \le \frac{1}{r}\left(1+\frac{\pi}{2}\right)|u-v|.$$

PROOF. We will first prove the estimate assuming $|u| = |v| = 1$, then $|u| = |v| = r$, and finally $|u|, |v| \ge r$. To this end, let $H^\epsilon = \eta^\epsilon * H$ where η^ϵ is a standard mollifier. That is, $\eta \in C_c^\infty(\mathbb{R}^3)$ is nonnegative, the support of η is the unit ball in $\mathbb{R}^3$, with $\int_{\mathbb{R}^3}\eta(y)dy = 1$, and $\eta^\epsilon(x) = \eta(x/\epsilon)/\epsilon^3$ for $\epsilon > 0$. It is not hard to see that H^ϵ is smooth and also converges in $C^1_{\text{loc}}(\mathbb{R}^3\setminus\{0\})$ to H (see Chapter 4 of [**21**] for example). Let $u, v \in \mathbb{S}^2$ and choose a geodesic path $\gamma : [0,1] \to \mathbb{S}^2$ joining u to v. Then

$$DH^\epsilon(v) - DH^\epsilon(u) = \int_0^1 D^2H^\epsilon(\gamma(t))\dot\gamma(t)dt.$$

Observe that for $\epsilon < 1$ and $w \in \mathbb{S}^2$,

$$\begin{aligned} D^2H^\epsilon(u)w\cdot w &= \int_{B_\epsilon(0)} \eta^\epsilon(x)D^2H(u-x)w\cdot w dx \\ &\le \int_{B_\epsilon(0)} \eta^\epsilon(x)\frac{1}{|u-x|}dx \\ &\le \int_{B_\epsilon(0)} \eta^\epsilon(x)\frac{1}{|u|-|x|}dx \\ &\le \int_{B_\epsilon(0)} \eta^\epsilon(x)\frac{1}{1-\epsilon}dx \\ &= \frac{1}{1-\epsilon}. \end{aligned}$$

It follows that

$$|DH^\epsilon(v) - DH^\epsilon(u)| \le \int_0^1 |D^2H^\epsilon(\gamma(t))||\dot\gamma(t)|dt \le \frac{1}{1-\epsilon}\int_0^1 |\dot\gamma(t)|dt = d(u,v)\frac{1}{1-\epsilon},$$

where

$$d(u,v) := \cos^{-1}(u\cdot v)$$

is the standard metric on $\mathbb{S}^2$. As

$$d(u,v) \le \frac{\pi}{2}|u-v|$$

for $u,v \in \mathbb{S}^2$, we may send $\epsilon \to 0$ and arrive at

$$|DH(u) - DH(v)| \le \frac{\pi}{2}|u-v|.$$

When $|u| = |v| = r$, we use the fact that DH is 0-degree homogeneous on $\mathbb{R}^3 \setminus \{0\}$ to get

$$|DH(u) - DH(v)| = |DH(u/r) - DH(v/r)| \le \frac{\pi}{2}|u/r - v/r| = \frac{\pi}{2r}|u-v|.$$

Now let $|v| \ge |u| \ge r$. By the triangle inequality

$$|DH(v) - DH(u)| \le \left|DH(v) - DH\left(|u|\frac{v}{|v|}\right)\right| + \left|DH\left(|u|\frac{v}{|v|}\right) - DH(u)\right|. \tag{2.5}$$

We can estimate the second term from the computation above:

$$\left|DH\left(|u|\frac{v}{|v|}\right) - DH(u)\right| \le \frac{\pi}{2r}\left|u - |u|\frac{v}{|v|}\right|.$$

An elementary computation shows

$$\left|u - |u|\frac{v}{|v|}\right| \le |u-v|;$$

in general, this inequality holds provided $|v| \ge |u|$ with $|v| > 0$. Therefore,

$$\left|DH\left(|u|\frac{v}{|v|}\right) - DH(u)\right| \le \frac{\pi}{2r}|u-v|.$$

As for the first term in (2.5), we employ the linear path

$$\alpha(t) = |u|\frac{v}{|v|} + t\left(v - |u|\frac{v}{|v|}\right), \quad 0 \le t \le 1.$$

Note in particular that

$$|\alpha(t)| \ge \alpha(t)\cdot\frac{v}{|v|} = |u| + t(|v| - |u|) \ge |u| \ge r.$$

Consequently, if H is smooth

$$\begin{aligned}\left|DH(v) - DH\left(|u|\frac{v}{|v|}\right)\right| &= |DH(\alpha(1)) - DH(\alpha(0))| \\ &\le \int_0^1 |D^2H(\alpha(t))||\dot\alpha(t)|dt \\ &\le \left|v - |u|\frac{v}{|v|}\right|\int_0^1 \frac{1}{|\alpha(t)|}dt \\ &\le \frac{|v| - |u|}{r} \\ &\le \frac{|u-v|}{r}.\end{aligned}$$

Otherwise, we can smooth H with a mollifier and derive the same inequality. Putting the two bounds together gives

$$|DH(u) - DH(v)| \le \frac{1}{r}\left(1 + \frac{\pi}{2}\right)|u-v|$$

for $|u|, |v| \geq r$. $\square$

COROLLARY 2.10. *Suppose for each $m \in \mathbb{N}$ that H^m is the support function of constant width body $K^m \subset \mathbb{R}^3$ with circumball centered at a^m. There is a subsequence $(H^{m_j})_{j\in\mathbb{N}}$ and a support function H^∞ of a constant width body $K^\infty \subset \mathbb{R}^3$ such that*

$$\lim_{j\to\infty} \max_{|u|\leq s} |H^{m_j}(u) - a^{m_j} \cdot u - H^\infty(u)| = 0 \tag{2.6}$$

and

$$\lim_{j\to\infty} \max_{r\leq|u|\leq 1/r} |DH^{m_j}(u) - a^{m_j} - DH^\infty(u)| = 0 \tag{2.7}$$

for each $s > 0$ and $0 < r \leq 1$.

PROOF. Set

$$\tilde{H}^m(u) := H^m(u) - a^{m_j} \cdot u$$

for $u \in \mathbb{R}^3$. By Corollary 2.8,

$$|\tilde{H}^m(u) - \tilde{H}^m(v)| \leq |u - v|$$

for all $u, v \in \mathbb{R}^3$ and each $m \in \mathbb{N}$. Since $H^m(0) = 0$ for each $m \in \mathbb{N}$, the Arzelà-Ascoli theorem implies $\tilde{H}^{m_j}$ has a locally uniformly convergent subsequence. The limit function H^∞ is necessarily continuous, sublinear, and satisfies the constant width condition (2.2). As a result, it is the support function of the constant width body

$$K^\infty := \bigcap_{|u|=1} \left\{x \in \mathbb{R}^3 : x \cdot u \leq H^\infty(u)\right\}.$$

This establishes the first limit (2.6). As for the second limit, we recall

$$|D\tilde{H}^m(u) - D\tilde{H}^m(v)| \leq \frac{1}{r}\left(1 + \frac{\pi}{2}\right)|u - v|$$

for $r \leq |u|, |v| \leq 1/r$ for each $0 < r \leq 1$. It follows that $(\tilde{H}^{m_j})_{j\in\mathbb{N}}$ converges in $C^1_{\text{loc}}(\mathbb{R}^3 \setminus \{0\})$ to H^∞. This establishes (2.7). $\square$

REMARK 2.11. The limit (2.6) implies that

$$K^{m_j} + a^{m_j} \to K^\infty$$

in the Hausdorff topology. This could have also been deduced as a consequence of Blaschke's selection theorem (Theorem 2.5.2 in [**37**]). Moreover, the argument above can be easily adapted to fashion an elementary proof of Blaschke's selection theorem.

2.3. Formula for the volume. Let us set

$$h := H|_{\mathbb{S}^2}$$

as the restriction of H to the unit sphere in $\mathbb{R}^3$. Since H is positively homogeneous, this function encodes all information about H. It also naturally inherits various properties from H such as $h \in C^{1,1}(\mathbb{S}^2)$ and

$$h(u) + h(-u) = 1 \tag{2.8}$$

for each $u \in \mathbb{S}^2$.

We can also infer information on the gradient and Hessian of $h : \mathbb{S}^2 \to \mathbb{R}$. The following assertion was proved by Howard along the way to his solution of Nakajima's problem in $\mathbb{R}^3$ [**27**].

LEMMA 2.12. (i) *For each* $u \in \mathbb{S}^2$,

$$DH(u) = \nabla h(u) + h(u)u \in \partial K$$

and u *is an outward normal vector to* ∂K *at* $DH(u)$.
(ii) *The map* $DH : \mathbb{S}^2 \to \partial K$ *is Lipschitz and surjective and its gradient is given by*

$$\nabla(DH)(u) = \nabla^2 h(u) + h(u)\mathrm{id}_{u^\perp}$$

for σ *almost every* $u \in \mathbb{S}^2$. *Here* $u^\perp := \{v \in \mathbb{R}^3 : v \cdot u = 0\}$.
(iii) *The linear transformation*

$$\nabla(DH)(u) : u^\perp \to u^\perp$$

is nonnegative definite for σ *almost every* $u \in \mathbb{S}^2$.

Thus far, much of what we've done in $\mathbb{R}^3$ carries over with little to no change in $\mathbb{R}^n$. The following result however is specific to convex bodies in $\mathbb{R}^3$. The expression below for the volume of K is known as *Blaschke's relation*.

PROPOSITION 2.13.

$$V(K) = \frac{1}{2}\sigma(\partial K) - \frac{\pi}{3}, \tag{2.9}$$

where

$$\sigma(\partial K) = \int_{\mathbb{S}^2} \left(h^2 - \frac{1}{2}|\nabla h|^2\right) d\sigma. \tag{2.10}$$

PROOF. 1. Let $t > 0$ and consider $K^t := K + B_t(0)$. This is a convex body with support function

$$H^t(u) := H(u) + t|u|.$$

Note that

$$DH^t(u) = DH(u) + tu = \nabla h(u) + h(u)u + tu$$

for $u \in \mathbb{S}^2$. It follows that

$$(DH^t(u) - DH^t(v)) \cdot (u - v) \geq t|u - v|^2$$

and thus

$$|DH^t(u) - DH^t(v)| \geq t|u - v|$$

for $u, v \in \mathbb{S}^2$. As a result, $DH^t : \mathbb{S}^2 \to \partial(K^t)$ is a bi-Lipschitz map.

Let $u^t : \partial(K^t) \to \mathbb{S}^2$ denote the inverse of DH^t; by part (i) of Lemma 2.12 this is the outward unit normal field on $\partial(K^t)$. Applying the divergence theorem, the area formula (Theorem 3.9 of [**21**]), and parts (*ii*) and (*iii*) of Lemma 2.12, we find

$$\begin{aligned} V(K^t) &= \int_{K^t} dx \\ &= \frac{1}{3}\int_{K^t} \operatorname{div}(x)dx \\ &= \frac{1}{3}\int_{\partial(K^t)} x\cdot u^t(x)d\sigma(x) \\ &= \frac{1}{3}\int_{\mathbb{S}^2} DH^t(u)\cdot u\ \det(\nabla(DH^t)(u))d\sigma(u) \\ &= \frac{1}{3}\int_{\mathbb{S}^2} H^t(u)\det(\nabla(DH^t)(u))d\sigma(u) \\ &= \frac{1}{3}\int_{\mathbb{S}^2} (h(u)+t)\det(\nabla^2 h(u) + (h(u)+t)\mathrm{id}_{u^\perp})d\sigma(u). \end{aligned}$$

Note that $K^s \subset K^t$ for $0 < s \le t$ and $K = \bigcap_{t>0} K^t$. We can then employ the monotonicity of Lebesgue measure and send $t \to 0^+$ to get

$$V(K) = \frac{1}{3}\int_{\mathbb{S}^2} h(u)\det(\nabla^2 h(u) + h(u)\mathrm{id}_{u^\perp})d\sigma(u).$$

2. Similarly we find

$$\sigma(\partial K^t) = \int_{\mathbb{S}^2} \det(\nabla^2 h(u) + (h(u)+t)\mathrm{id}_{u^\perp})d\sigma(u) \tag{2.11}$$

for $t > 0$. We claim this formula also holds at $t = 0$. To see this, we let $r > 0$ be the inradius of K and observe

$$K = (1-t)K + tK \supset (1-t)K + tB_r(0) = (1-t)[K^{tr/(1-t)}]$$

for $t \in (0,1)$. Since $\sigma(\partial A) \le \sigma(\partial B)$ for two convex bodies with $A \subset B$ [**51**],

$$\sigma(K) \ge (1-t)^2\sigma(\partial K^{tr/(1-t)}) = (1-t)^2\int_{\mathbb{S}^2}\det\left(\nabla^2 h(u) + \left(h(u) + \frac{rt}{1-t}\right)\mathrm{id}_{u^\perp}\right)d\sigma(u)$$

for $t \in (0,1)$. Likewise, $\sigma(\partial K)$ is bounded from above by the right hand side of (2.11) for each $t > 0$. We conclude

$$\sigma(\partial K) = \int_{\mathbb{S}^2} \det(\nabla^2 h(u) + h(u)\mathrm{id}_{u^\perp})d\sigma(u).$$

We will now employ the coarea formula (Theorem 3.11 of [**21**]) with the Lipschitz function

$$f(x) = d(x, K).$$

Note that $|Df(x)| = 1$ almost everywhere in K^c. In particular, for $t \geq 0$

$$\begin{aligned} V(K^t) - V(K) &= \int_{K^t \setminus K} dx \\ &= \int_{\mathbb{R}^3} 1_{K^t \setminus K}(x) dx \\ &= \int_{\mathbb{R}^3} 1_{K^t \setminus K}(x) |Df(x)| dx \\ &= \int_0^\infty \left\{ \int_{f^{-1}(\{s\})} 1_{K^t \setminus K}(x) d\sigma(x) \right\} ds \\ &= \int_0^\infty \sigma((K^t \setminus K) \cap f^{-1}(\{s\}) ds \\ &= \int_0^t \sigma(\partial K^s) ds. \end{aligned}$$

Here we used $d(x, K) = s > 0$ if and only if $x \in \partial K^s$. Thus,

$$\begin{aligned} \sigma(\partial K) &= \frac{d}{dt} V(K^t) \Big|_{t=0} \\ &= \frac{d}{dt} \frac{1}{3} \int_{\mathbb{S}^2} (h(u) + t) \det(\nabla^2 h(u) + (h(u) + t) \mathrm{id}_{u^\perp}) d\sigma(u) \Big|_{t=0} \\ &= \frac{1}{3} \int_{\mathbb{S}^2} \det(\nabla^2 h(u) + h(u) \mathrm{id}_{u^\perp}) d\sigma(u) + \frac{1}{3} \int_{\mathbb{S}^2} h(u)(\Delta h(u) + 2h(u)) d\sigma(u) \\ &= \frac{1}{3} \sigma(\partial K) + \frac{1}{3} \int_{\mathbb{S}^2} h(u)(\Delta h(u) + 2h(u)) d\sigma(u). \end{aligned}$$

That is,

$$\begin{aligned} \sigma(\partial K) &= \frac{1}{2} \int_{\mathbb{S}^2} h(\Delta h + 2h) d\sigma \\ &= \frac{1}{2} \int_{\mathbb{S}^2} h\Delta h + 2h^2 d\sigma \\ &= \int_{\mathbb{S}^2} \left(h^2 - \frac{1}{2} |\nabla h|^2 \right) d\sigma, \end{aligned}$$

which is (2.10).

3. As for Blaschke's relation, we will use the fact that Lebesgue measure is invariant under orthogonal transformations so that $V(K) = V(-K)$. Moreover, the support function of $-K$ is

$$H(-u) = |u| - H(u) \quad (u \in \mathbb{R}^3),$$

as K has constant width. Consequently,

$$
\begin{aligned}
&V(K)\\
&= V(-K)\\
&= \frac{1}{3}\int_{\mathbb{S}^2} h(-u)\det(\nabla^2 h(-u) + h(-u)\mathrm{id}_{u^\perp})d\sigma(u)\\
&= \frac{1}{3}\int_{\mathbb{S}^2} (1-h(u))\det(-\nabla^2 h(u) + (1-h(u))\mathrm{id}_{u^\perp})d\sigma(u)\\
&= \frac{1}{3}\int_{\mathbb{S}^2} (1-h)\det(1-(\nabla^2 h + h\mathrm{id}_{u^\perp}))d\sigma\\
&= \frac{1}{3}\int_{\mathbb{S}^2} (1-h)[1-(\Delta h + 2h) + \det(\nabla^2 h + h\mathrm{id}_{u^\perp})]d\sigma\\
&= \frac{1}{3}\int_{\mathbb{S}^2} [1-(\Delta h + 2h) + \det(\nabla^2 h + h\mathrm{id}_{u^\perp}) - h(1-(\Delta h + 2h) + \det(\nabla^2 h + h\mathrm{id}_{u^\perp}))]d\sigma\\
&= \frac{1}{3}\int_{\mathbb{S}^2} \left[1-(\Delta h + 2h) + \det(\nabla^2 h + h\mathrm{id}_{u^\perp}) - h + h(\Delta h + 2h) - h\det(\nabla^2 h + h\mathrm{id}_{u^\perp})\right] d\sigma.
\end{aligned}
$$

In view of the constant width condition (2.8),

$$\frac{1}{4\pi}\int_{\mathbb{S}^2} h d\sigma = \frac{1}{2}.$$

Since $\Delta h = \mathrm{div}(\nabla h)$ integrates to 0 over $\mathbb{S}^2$, it follows that

$$\int_{\mathbb{S}^2} [1-(\Delta h + 2h)]d\sigma = 0.$$

Using the formulae above that we derived for $V(K)$ and $\sigma(\partial K)$ then gives

$$
\begin{aligned}
&V(K)\\
&= \frac{1}{3}\int_{\mathbb{S}^2} \left[1-(\Delta h + 2h) + \det(\nabla^2 h + h\mathrm{id}_{u^\perp}) - h + h(\Delta h + 2h) - h\det(\nabla^2 h + h\mathrm{id}_{u^\perp})\right] d\sigma\\
&= \frac{1}{3}\sigma(\partial K) - \frac{2\pi}{3} + \frac{1}{3}\int_{\mathbb{S}^2} h(\Delta h + 2h)d\sigma - V(K)\\
&= \sigma(\partial K) - \frac{2\pi}{3} - V(K).
\end{aligned}
$$

We conclude (2.9). □

3. Spaces of functions and measures

In this section, we will study the various spaces that will be needed in our analysis of the doubly nonlinear evolution (1.4). Some of these spaces were introduced in subsection 1.2. First, we will show that the space $\mathcal{C}$ is compact in a certain sense. Then we will consider a subspace of $H^1(\mathbb{S}^2)$ with an inner product tailored for this work. Next, we will study measures in $\mathcal{P}^\perp$ modulo the ones which vanish on $\mathcal{C}$; we'll also identify an important compact subset of this quotient space that will play a crucial role in our subsequent existence proof. Finally, we will introduce the notion of the E^* variation of a path $\xi : [0,\infty) \to \mathcal{P}^\perp$.

3.1. The space $\mathcal{C}$. Let us recall the space $\mathcal{C}$ defined in (1.2). We will argue that this space is convex and inherits the compactness we discussed above for support functions.

LEMMA 3.1. (i) *If $g \in \mathcal{C}$, then*

$$\|\mathbf{g}\| \le \sqrt{\frac{3}{8}} - \frac{1}{2}.$$

(ii) *$\mathcal{C}$ is convex.*
(iii) *For any sequence $(g^k)_{k\in\mathbb{N}} \subset \mathcal{C}$, there is a subsequence $(\mathbf{g}^{k_j})_{j\in\mathbb{N}}$ which converges to some $\mathbf{g}$ with $g \in \mathcal{C}$. Moreover, there is $a^j \in \mathbb{R}^3$ for each $j \in \mathbb{N}$ such that $g^{k_j}(u) - a^j \cdot u$ converges in $C^1(\mathbb{S}^2)$ to g.*

PROOF. (i) Suppose $g \in \mathcal{C}$. There is a constant width body such that $g + 1/2$ is the $\mathbb{S}^2$ restriction of the support function of $K \subset \mathbb{R}^3$. By Corollary 2.5,

$$R(K) = \frac{1}{2} + \|\mathbf{g}\| \le \sqrt{\frac{3}{8}}.$$

(ii) Assume $g_0, g_1 \in \mathcal{C}$ and let H_0, H_1 denote the corresponding support functions with respective constant width bodies $K_0, K_1 \subset \mathbb{R}^3$. Suppose $\lambda \in [0,1]$ and note that the support function of the convex body $(1-\lambda)K_0+\lambda K_1$ is $(1-\lambda)H_0+\lambda H_1$. Since

$$(1-\lambda)H_0(u) + \lambda H_1(u) + [(1-\lambda)H_0(-u) + \lambda H_1(-u)] = |u|,$$

$(1-\lambda)K_0 + \lambda K_1$ has constant width. As

$$((1-\lambda)g_0 + \lambda g_1) + 1/2 = (1-\lambda)(g_0 + 1/2) + \lambda(g_1 + 1/2) = [(1-\lambda)H_0 + \lambda H_1]|_{\mathbb{S}^2},$$

we conclude $(1-\lambda)g_0 + \lambda g_1 \in \mathcal{C}$. Therefore, $\mathcal{C}$ is convex.

(iii) Let

$$g^k + 1/2 = H^k|_{\mathbb{S}^2},$$

where H^k is the support function of a constant width body. By Corollary 2.10, there are $a^j \in \mathbb{R}^3$ and a subsequence $(H^{k_j})_{j\in\mathbb{N}}$ for which $H^{k_j}(u) - a^j \cdot u$ converges uniformly to the support function of a constant width body H on compact subsets of $\mathbb{R}^3$. If $g + 1/2 = H|_{\mathbb{S}^2}$,

$$\|\mathbf{g}^{k_j} - \mathbf{g}\| \le \max_{|u|=1} |g^{k_j}(u) - g(u) - a^j \cdot u| = \max_{|u|=1} |H^{k_j}(u) - H(u) - a^j \cdot u| \to 0$$

as $j \to \infty$.

Corollary 2.10 also gives that $DH^{k_j}(u) - a^j$ converges to $DH(u)$ uniformly for $u \in \mathbb{S}^2$. This implies

$$\nabla(g^{k_j}(u)-a^j\cdot u) = DH^{k_j}(u)-a^j-(g^{k_j}(u)-a^j\cdot u+1/2)u \to DH(u)-(g(u)+1/2)u = \nabla g(u)$$

uniformly for $u \in \mathbb{S}^2$. □

3.2. Square integrable paths. The Sobolev space $H^1(\mathbb{S}^2)$ is the completion of $C^\infty(\mathbb{S}^2)$ in the norm

$$g \mapsto \left(\int_{\mathbb{S}^2} (g^2 + |\nabla g|^2)\, d\sigma\right)^{1/2}.$$

We refer the reader to Chapter 2 of [**26**] for more on $H^1(\mathbb{S}^2)$. We will consider the closed subspace

$$\mathcal{V} := \left\{g \in H^1(\mathbb{S}^2) : \int_{\mathbb{S}^2} g d\sigma = 0\right\}$$

which admits the semi-inner product

$$(g,h) := \int_{\mathbb{S}^2} (\nabla g \cdot \nabla h - 2gh) d\sigma.$$

We note that if $g \in \mathcal{V}$,

$$(g, g) = \int_{\mathbb{S}^2} (|\nabla g|^2 - 2g^2) d\sigma$$

is nonnegative and vanishes if and only if $g \in \mathcal{P}$. We will quantify this in the lemma below. Thus, it is natural to consider the quotient space

$$\mathcal{V}/\mathcal{P}$$

with quotient inner product defined as

$$(\mathbf{g}, \boldsymbol{h}) := (g, h).$$

Here we are extending our notation $\mathbf{g} = g + \mathcal{P}$ to $g \in \mathcal{V}$.

We claim that $\mathcal{V}/\mathcal{P}$ is a Hilbert space. First, we'll derive a basic stability estimate.

LEMMA 3.2. *For each $g \in \mathcal{V}$, the following inequality holds*

$$4 \min_{a \in \mathbb{R}^3} \int_{\mathbb{S}^2} (g(u) - a \cdot u)^2 d\sigma(u) \leq \int_{\mathbb{S}^2} (|\nabla g|^2 - 2g^2) d\sigma. \tag{3.1}$$

PROOF. Recall the collection of spherical harmonics ϕ_ℓ^m, where $\ell = 0, 1, 2, \dots$ and $m = -\ell, -\ell + 1, \dots, \ell - 1, \ell$. These are eigenfunctions for $-\Delta$ which form an orthonormal basis for $L^2(\mathbb{S}^2)$. That is,

$$\int_{\mathbb{S}^2} \phi_\ell^m \phi_{\ell'}^{m'} d\sigma = \delta_{mm'} \delta_{\ell\ell'}$$

and

$$-\Delta \phi_\ell^m = \ell(\ell + 1) \phi_\ell^m$$

in $\mathbb{S}^2$ for each $\ell, \ell' = 0, 1, 2, \dots$, $m = -\ell, -\ell + 1, \dots, \ell - 1, \ell$, and $m' = -\ell', -\ell' + 1, \dots, \ell' - 1, \ell'$. We especially note that ϕ_0^0 is a nonzero constant and we may assume

$$\phi_{-1}^1(u) = cu_1, \quad \phi_0^1(u) = cu_2, \quad \phi_1^1(u) = cu_3.$$

Here $c = \sqrt{3/(4\pi)}$ is the normalization constant to ensure these functions have $L^2(\mathbb{S}^2)$ norm equal to 1.

As a result, we may write

$$g = \sum_{\ell=0}^{\infty} \sum_{m=-\ell}^{\ell} c_\ell^m \phi_\ell^m$$

where

$$c_\ell^m = \int_{\mathbb{S}^2} g \phi_\ell^m d\sigma.$$

We note that $c_0^0 = 0$ since the average of g is 0. It follows from these observations that

$$\int_{\mathbb{S}^2} g^2 d\sigma = \sum_{\ell=1}^{\infty} \sum_{m=-\ell}^{\ell} (c_\ell^m)^2$$

and

$$\int_{\mathbb{S}^2} |\nabla g|^2 d\sigma = \sum_{\ell=1}^{\infty} \sum_{m=-\ell}^{\ell} \ell(\ell + 1)(c_\ell^m)^2.$$

This leads to

$$\begin{aligned}\int_{\mathbb{S}^2}\left(|\nabla g|^2-2g^2\right)d\sigma &= \sum_{\ell=1}^{\infty}\sum_{m=-\ell}^{\ell}[\ell(\ell+1)-2](c_\ell^m)^2\\ &= \sum_{\ell=2}^{\infty}\sum_{m=-\ell}^{\ell}[\ell(\ell+1)-2](c_\ell^m)^2\\ &\ge 4\sum_{\ell=2}^{\infty}\sum_{m=-\ell}^{\ell}(c_\ell^m)^2\\ &= 4\int_{\mathbb{S}^2}(g(u)-c(c_{-1}^1u_1+c_0^1u_2+c_1^1u_3))^2d\sigma(u)\\ &= 4\min_{a\in\mathbb{R}^3}\int_{\mathbb{S}^2}(g(u)-a\cdot u)^2d\sigma(u).\end{aligned}$$

□

COROLLARY 3.3. *$\mathcal{V}/\mathcal{P}$ endowed with the inner product $(\cdot,\cdot)$ is complete.*

PROOF. Suppose $(\mathbf{g}^k)_{k\in\mathbb{N}}\subset\mathcal{V}/\mathcal{P}$ is a Cauchy sequence. For each $g^k\in\mathcal{V}$, we select $a^k\in\mathbb{R}^3$ to satisfy

$$\min_{a\in\mathbb{R}^3}\int_{\mathbb{S}^2}(g^k(u)-a\cdot u)^2d\sigma(u)=\int_{\mathbb{S}^2}(g^k(u)-a^k\cdot u)^2d\sigma(u).$$

We also set $\tilde{g}^k(u)=g^k(u)-a^k\cdot u$ and note $\tilde{g}^k\in\mathbf{g}^k$ for each $k\in\mathbb{N}$. In view of the previous lemma, for each $\epsilon>0$, there is $N\in\mathbb{N}$ so that

$$\epsilon\ge\int_{\mathbb{S}^2}(|\nabla(\tilde{g}^k-\tilde{g}^\ell)|^2-2(\tilde{g}^k-\tilde{g}^\ell)^2)d\sigma\ge 4\int_{\mathbb{S}^2}(\tilde{g}^k-\tilde{g}^\ell)^2d\sigma$$

for each $k,\ell\ge N$. As a result, $(\tilde{g}^k)_{k\in\mathbb{N}}$ is Cauchy in $L^2(\mathbb{S}^2)$. It also follows from the above inequality that $(\tilde{g}^k)_{k\in\mathbb{N}}$ is Cauchy in $H^1(\mathbb{S}^2)$. Since $H^1(\mathbb{S}^2)$ is complete and $\mathcal{V}$ is closed, there is $g\in\mathcal{V}$ such that $\tilde{g}^k\to g$ in $H^1(\mathbb{S}^2)$. We conclude

$$(\mathbf{g}^k-\mathbf{g},\mathbf{g}^k-\mathbf{g})=\int_{\mathbb{S}^2}(|\nabla(\tilde{g}^k-g)|^2-2(\tilde{g}^k-g)^2)d\sigma\to 0$$

as $k\to\infty$. □

Suppose $\mathbf{g}:[0,\infty)\to\mathcal{V}/\mathcal{P}$ is a measurable mapping. We may consider the integral

$$\int_0^\infty(\mathbf{g}(t),\mathbf{g}(t))dt=\int_0^\infty\int_{\mathbb{S}^2}\left(|\nabla g(t)|^2-2g(t)^2\right)d\sigma dt$$

which may or may not be finite. Whenever this integral is finite, we say that

$$\mathbf{g}\in L^2([0,\infty);\mathcal{V}/\mathcal{P}).$$

Further, we may view this space as a Hilbert space with the inner product

$$(\mathbf{g}_1,\mathbf{g}_2)\mapsto\int_0^\infty(\mathbf{g}_1(t),\mathbf{g}_2(t))dt$$

if we identify paths that are equal almost everywhere.

A basic assertion that we will make use of is as follows.

LEMMA 3.4. *The collection of* $\mathbf{g} \in L^2([0,\infty); \mathcal{V}/\mathcal{P}); t \mapsto \mathbf{g}(t)$ *such that*

$$g(t) \in \mathcal{C}$$

for almost every $t \geq 0$ *is weakly closed.*

PROOF. Let us call the collection of maps in question $\mathcal{K}$. We claim that $\mathcal{K}$ is convex and closed. The result would then follow from Mazur's theorem (Chapter V of [**54**]). The convexity of $\mathcal{K}$ follows easily since $\mathcal{C}$ is convex. Now suppose $(\mathbf{g}^k)_{k\in\mathbb{N}} \subset \mathcal{K}$ converges in $L^2([0,\infty); \mathcal{V}/\mathcal{P})$ to some $\mathbf{g}$. We need to verify $\mathbf{g} \in \mathcal{K}$.

To this end, we subtract a subsequence $(\mathbf{g}^{k_j})_{j\in\mathbb{N}}$ such that $\mathbf{g}^{k_j}(t) \to \mathbf{g}(t)$ for almost every $t \geq 0$ in $\mathcal{V}/\mathcal{P}$. For any such t, Lemma 3.1 implies the existence of a sequence $(g^j(t))_{j\in\mathbb{N}} \subset \mathcal{C}$ with $g^j(t) \in \mathbf{g}^{k_j}(t)$ which has a subsequence that converges in $C^1(\mathbb{S}^2)$ to some limit function $\tilde{g}(t) \in \mathcal{C}$. It follows that along an appropriate subsequence $(g^j(t), h)$ converges to both $(g(t), h)$ and $(\tilde{g}(t), h)$ for $\boldsymbol{h} \in \mathcal{V}/\mathcal{P}$. That is,

$$(g(t), h) = (\tilde{g}(t), h)$$

for all $\boldsymbol{h} \in \mathcal{V}/\mathcal{P}$. As a result, $g(t) - \tilde{g}(t) \in \mathcal{P}$ which implies $g(t) \in \mathcal{C}$. We conclude that for almost every $t \geq 0$, $g(t) \in \mathcal{C}$. Thus, $\mathcal{K}$ is closed. □

The following claim will also be useful for us.

LEMMA 3.5. *Suppose* $\mathbf{g} \in L^2([0,\infty); \mathcal{V}/\mathcal{P}); t \mapsto \mathbf{g}(t)$ *satisfies*

$$g(t) \in \mathcal{C}$$

for almost every $t \geq 0$. *There is a measurable*

$$\tilde{\mathbf{g}} : [0,\infty) \to C(\mathbb{S}^2)/\mathcal{P}; t \mapsto \tilde{\mathbf{g}}(t)$$

with $\tilde{g}(t) \in \mathcal{C}$ *for* $t \geq 0$ *and*

$$\tilde{\mathbf{g}}(t) = \mathbf{g}(t)$$

for almost every $t \geq 0$.

PROOF. Let $N \subset [0,\infty)$ be a null set for which $g(t) \in \mathcal{C}$ for $t \notin N$. Set

$$\tilde{\mathbf{g}}(t) := \begin{cases} \mathbf{g}(t), & t \notin N \\ 0, & t \in N. \end{cases}$$

Then $\tilde{g}(t) \in \mathcal{C}$ for each $t \geq 0$ and $\tilde{\mathbf{g}}(t) = \mathbf{g}(t)$ for almost every $t \geq 0$.

Now assume $\mu \in \mathcal{P}^\perp$. According to Proposition A.2 in the appendix: for each $\epsilon \in (0,1)$, there is $\mu^\epsilon \in C^\infty(\mathbb{S}^2)$ such that

$$\int_{\mathbb{S}^2} u_i \mu^\epsilon d\sigma = 0 \tag{3.2}$$

for $i = 1, 2, 3$, and

$$\left| \int_{\mathbb{S}^2} g(t)\mu^\epsilon d\sigma - \int_{\mathbb{S}^2} g(t) d\mu \right| \leq 4\sqrt{2\epsilon}\, \|\mu\|_*$$

for all $t \notin N$. Therefore,

$$\int_{\mathbb{S}^2} g(t) d\mu = \lim_{\epsilon \to 0^+} \int_{\mathbb{S}^2} g(t)\mu^\epsilon d\sigma$$

for $t \notin N$.

As a result, it suffices to show

$$[0,\infty) \ni t \mapsto \int_{\mathbb{S}^2} g(t)\mu^\epsilon d\sigma \tag{3.3}$$

is measurable as

$$[0,\infty) \ni t \mapsto \int_{\mathbb{S}^2} g(t) d\mu$$

would be the almost everywhere limit of measurable functions. Pettis' theorem (Chapter V section 4 of [**54**]) would then imply that $\tilde{\mathbf{g}} : [0,\infty) \to C(\mathbb{S}^2)/\mathcal{P}$ is measurable.

We now focus on showing (3.3) is measurable. To this end, we fix an element $\boldsymbol{h} \in \mathcal{V}/\mathcal{P}$. In view of (3.2), we have

$$\begin{aligned}\int_{\mathbb{S}^2} \mu^\epsilon(u)h(u)d\sigma(u) &= \int_{\mathbb{S}^2} \mu^\epsilon(u)(h(u) - a\cdot u)d\sigma(u)\\ &\le \left(\int_{\mathbb{S}^2}(\mu^\epsilon)^2 d\sigma\right)^{1/2}\left(\int_{\mathbb{S}^2}(h(u)-a\cdot u)^2 d\sigma(u)\right)^{1/2}\end{aligned}$$

for any $a \in \mathbb{R}^3$. Employing (3.1), we additionally find

$$\begin{aligned}\int_{\mathbb{S}^2} \mu^\epsilon h d\sigma &\le \left(\int_{\mathbb{S}^2}(\mu^\epsilon)^2 d\sigma\right)^{1/2}\frac{1}{2}\left(\int_{\mathbb{S}^2}\left(|\nabla h|^2 - 2h^2\right)d\sigma\right)^{1/2}\\ &= \frac{1}{2}\left(\int_{\mathbb{S}^2}(\mu^\epsilon)^2 d\sigma\right)^{1/2}(\boldsymbol{h},\boldsymbol{h})^{1/2}.\end{aligned}$$

Therefore, the linear functional

$$\mathcal{V}/\mathcal{P} \ni \boldsymbol{h} \mapsto \int_{\mathbb{S}^2} \mu^\epsilon h d\sigma$$

is continuous. Since $\mathbf{g} : [0,\infty) \to \mathcal{V}/\mathcal{P}$ is measurable, it follows that (3.3) is measurable. □

3.3. The metric induced by $\chi^*_{\mathcal{C}}$.

Let us define

$$\mathcal{C}^\perp := \{\xi \in M(\mathbb{S}^2) : \xi|_{\mathcal{C}} = 0\},$$

and observe that $\mathcal{C}^\perp \subset \mathcal{P}^\perp$ is a closed subspace. We will also consider the quotient space

$$\mathcal{P}^\perp/\mathcal{C}^\perp = \{\xi + \mathcal{C}^\perp : \xi \in \mathcal{P}^\perp\}.$$

As we did to ease notation when expressing elements of the quotient space $C(\mathbb{S}^2)/\mathcal{P}$, we will write

$$\boldsymbol{\xi} = \xi + \mathcal{C}^\perp$$

whenever $\xi \in \boldsymbol{\xi}$.

While $\mathcal{P}^\perp/\mathcal{C}^\perp$ admits the standard quotient norm

$$\|\boldsymbol{\xi}\|_* := \inf_{\eta\in\mathcal{C}^\perp}\|\xi+\eta\|_*,$$

is it also is endowed with another norm

$$\chi^*_{\mathcal{C}}(\boldsymbol{\xi}) := \sup\left\{\langle \xi, g\rangle : g \in \mathcal{C}\right\}. \tag{3.4}$$

We have labeled this norm $\chi^*_{\mathcal{C}}$ as it arises as the convex dual of the characteristic function

$$\chi_{\mathcal{C}}(g) := \begin{cases} 0, & g \in \mathcal{C} \\ +\infty, & g \notin \mathcal{C}. \end{cases}$$

Indeed the right hand side of (3.4) is the dual of $\chi_{\mathcal{C}}$; since this function is invariant under translations by elements of $\mathcal{C}^\perp$, we naturally consider it as a function on $\mathcal{P}^\perp/\mathcal{C}^\perp$.

First we note that the quotient norm on $\mathcal{P}^\perp/\mathcal{C}^\perp$ controls the norm $\chi^*_{\mathcal{C}}$. In particular, the topology associated with $\chi^*_{\mathcal{C}}$ is weaker than the topology determined by the quotient norm.

PROPOSITION 3.6. *For $\boldsymbol{\xi} \in \mathcal{P}^\perp/\mathcal{C}^\perp$,*

$$\chi^*_{\mathcal{C}}(\boldsymbol{\xi}) \le \left(\sqrt{\frac{3}{8}} - \frac{1}{2}\right) \|\boldsymbol{\xi}\|_*. \tag{3.5}$$

PROOF. We recall that if $g \in \mathcal{C}$, then $\|\mathbf{g}\| \le \sqrt{\frac{3}{8}} - \frac{1}{2}$. Therefore,

$$\begin{aligned} \chi^*_{\mathcal{C}}(\boldsymbol{\xi}) &= \sup\{\langle \xi, g\rangle : g \in \mathcal{C}\} \\ &\le \sup\left\{\langle \xi, g\rangle : \|\mathbf{g}\| \le \sqrt{\frac{3}{8}} - \frac{1}{2}\right\} \\ &= \left(\sqrt{\frac{3}{8}} - \frac{1}{2}\right)\|\boldsymbol{\xi}\|_*. \end{aligned}$$

□

Next, we will argue that quotient norm bounded subsets of $\mathcal{P}^\perp/\mathcal{C}^\perp$ are compact metric spaces when endowed with the metric

$$(\boldsymbol{\xi}, \boldsymbol{\zeta}) \mapsto \chi^*_{\mathcal{C}}(\boldsymbol{\xi} - \boldsymbol{\zeta})$$

induced by $\chi^*_{\mathcal{C}}$.

PROPOSITION 3.7. *For each $r > 0$,*

$$\mathcal{B}_r := \{\boldsymbol{\xi} \in \mathcal{P}^\perp/\mathcal{C}^\perp : \|\boldsymbol{\xi}\|_* \le r\}$$

is a compact metric space when endowed with metric induced by $\chi^*_{\mathcal{C}}$.

PROOF. Suppose that $(\boldsymbol{\xi}^k)_{k\in\mathbb{N}} \subset \mathcal{B}_r$. For each $k \in \mathbb{N}$, we may choose $\eta^k \in \mathcal{C}^\perp$ such that

$$r \ge \|\boldsymbol{\xi}^k\|_* \ge \|\xi^k + \eta^k\|_* - \frac{r}{2}.$$

Let us set $\zeta^k := \xi^k + \eta^k$ so that

$$\|\zeta^k\|_* \le 3r/2.$$

By Alaoglu's theorem, that there is a subsequence $(\zeta^{k_j})_{j\in\mathbb{N}}$ that converges weak* in $M(\mathbb{S}^2)$ to some $\xi \in \mathcal{P}^\perp$.

A routine application of Lemma 3.1 implies that for each $j \in \mathbb{N}$, there is $g^j \in \mathcal{C}$ with

$$\chi^*_{\mathcal{C}}(\boldsymbol{\xi}^{k_j} - \boldsymbol{\xi}) = \langle \xi^{k_j} - \xi, g^j\rangle = \langle \zeta^{k_j} - \xi, g^j\rangle.$$

We may also assume without any loss of generality that g^j converges uniformly to some $g \in \mathcal{C}$. Notice that

$$\begin{aligned}\chi_{\mathcal{C}}^*(\boldsymbol{\xi}^{k_j} - \boldsymbol{\xi}) &= \langle \zeta^{k_j} - \xi, g^j \rangle \\ &= \langle \zeta^{k_j} - \xi, g^j - g \rangle + \langle \zeta^{k_j} - \xi, g \rangle \\ &\le \|\zeta^{k_j} - \xi\|_* \|g^j - g\| + \langle \zeta^{k_j} - \xi, g \rangle \\ &\le 3r\|g^j - g\| + \langle \zeta^{k_j} - \xi, g \rangle.\end{aligned}$$

Therefore,

$$\limsup_{j\to\infty} \chi_{\mathcal{C}}^*(\boldsymbol{\xi}^{k_j} - \boldsymbol{\xi}) = 0.$$

□

COROLLARY 3.8. *Suppose* $(\boldsymbol{\xi}^k)_{k\in\mathbb{N}} \subset \mathcal{P}^\perp/\mathcal{C}^\perp$ *and* $\boldsymbol{\xi} \in \mathcal{P}^\perp/\mathcal{C}^\perp$ *with*

$$\lim_{k\to\infty} \chi_{\mathcal{C}}^*(\boldsymbol{\xi}^k - \boldsymbol{\xi}) = 0. \tag{3.6}$$

Then

$$\|\boldsymbol{\xi}\|_* \le \liminf_{k\to\infty} \|\boldsymbol{\xi}^k\|_*.$$

PROOF. Without any loss of generality, suppose

$$\liminf_{k\to\infty} \|\boldsymbol{\xi}^k\|_* = \lim_{j\to\infty} \|\boldsymbol{\xi}^{k_j}\|_* := L < \infty.$$

For a given $\epsilon > 0$, we can choose N such that

$$\|\boldsymbol{\xi}^{k_j}\|_* \le L + \epsilon$$

for $j \ge N$. By the previous proposition,

$$\{\boldsymbol{\zeta} \in \mathcal{P}^\perp/\mathcal{C}^\perp : \|\boldsymbol{\zeta}\|_* \le L + \epsilon\}$$

is a compact metric space when endowed with the metric induced by $\chi_{\mathcal{C}}^*$. By assumption (3.6), it must be that

$$\|\boldsymbol{\xi}\|_* \le L + \epsilon.$$

We conclude since $\epsilon > 0$ is arbitrary. □

3.4. E^* variation. Recall that the convex conjugate of E is given by

$$E^*(\zeta) = \sup\{\langle \zeta, g \rangle - E(g) : g \in \mathcal{C}\}$$

for $\zeta \in \mathcal{P}^\perp$. Note that E^* is convex, proper, and weak* lower-semicontinuous. Moreover, For a given path $\xi : [0,\infty) \to \mathcal{P}^\perp$, we will consider the integrals

$$\int_s^t E^*(\dot{\xi}(\tau))d\tau$$

for $0 \le s \le t < \infty$. However, we do not always want to require the almost everywhere weak* differentiability of ξ. This leads us to the following definition.

DEFINITION 3.9. Let $\xi : [0,\infty) \to \mathcal{P}^\perp$ and $0 \le s < t < \infty$. The E^* *variation* of ξ on $[s,t]$ is defined as

$$E^*V(\xi, s, t) := \sup\left\{\sum_{k=1}^N (\tau_k - \tau_{k-1}) E^*\left(\frac{\xi(\tau_k) - \xi(\tau_{k-1})}{\tau_k - \tau_{k-1}}\right) : s = \tau_0 < \cdots < \tau_N = t\right\}.$$

When $E^*V(\xi, s, t) < \infty$, ξ has *finite E^* variation* on $[s, t]$; if this is the case for all $[s, t] \subset [0, \infty)$, ξ has *locally finite E^* variation.* The *E^* variation* of ξ on $[0, \infty)$ is defined as

$$E^*V(\xi, 0, \infty) := \lim_{t\to\infty} E^*V(\xi, 0, t)$$

provided this limit is finite; in this case, we say ξ has *finite E^* variation.*

An elementary identity is as follows.

LEMMA 3.10. *Suppose $\xi : [0, \infty) \to \mathcal{P}^\perp$ has locally finite E^* variation. Then*

$$E^*V(\xi, s, t) = E^*V(\xi, 0, t) - E^*V(\xi, 0, s) \tag{3.7}$$

for all $0 \le s < t < \infty$.

PROOF. Let $\epsilon > 0$, choose a partition $0 = \tau_0 < \cdots < \tau_N = s$ such that

$$E^*V(\xi, 0, s) \le \sum_{k=1}^{N} (\tau_k - \tau_{k-1}) E^* \left(\frac{\xi(\tau_k) - \xi(\tau_{k-1})}{\tau_k - \tau_{k-1}} \right) + \frac{\epsilon}{2},$$

and select another partition $s = \tau_N < \cdots < \tau_M = t$ such that

$$E^*V(\xi, s, t) \le \sum_{k=N+1}^{M} (\tau_k - \tau_{k-1}) E^* \left(\frac{\xi(\tau_k) - \xi(\tau_{k-1})}{\tau_k - \tau_{k-1}} \right) + \frac{\epsilon}{2}.$$

As $0 = \tau_0 < \cdots < \tau_M = t$,

$$E^*V(\xi, 0, s) + E^*V(\xi, s, t) - \epsilon \le \sum_{k=1}^{M} (\tau_k - \tau_{k-1}) E^* \left(\frac{\xi(\tau_k) - \xi(\tau_{k-1})}{\tau_k - \tau_{k-1}} \right) \le E^*V(\xi, 0, t).$$

Therefore, $E^*V(\xi, 0, s) + E^*V(\xi, s, t) \le E^*V(\xi, 0, t)$.

Again let $\epsilon > 0$ and select any partition $0 = \tau_0 < \cdots < \tau_N = t$ of $[0, t]$ with

$$E^*V(\xi, 0, t) \le \sum_{j=1}^{N} (\tau_j - \tau_{j-1}) E^* \left(\frac{\xi(\tau_j) - \xi(\tau_{j-1})}{\tau_j - \tau_{j-1}} \right) + \epsilon.$$

Suppose there is some $1 \le k \le N$ for which $\tau_{k-1} < s < \tau_k$. Since

$$\frac{\xi(\tau_k) - \xi(\tau_{k-1})}{\tau_k - \tau_{k-1}} = \frac{s - \tau_{k-1}}{\tau_k - \tau_{k-1}} \frac{\xi(s) - \xi(\tau_{k-1})}{s - \tau_{k-1}} + \frac{\tau_k - s}{\tau_k - \tau_{k-1}} \frac{\xi(\tau_k) - \xi(s)}{\tau_k - s}$$

and E^* is convex,

$$E^* \left(\frac{\xi(\tau_k) - \xi(\tau_{k-1})}{\tau_k - \tau_{k-1}} \right) \le \frac{s - \tau_{k-1}}{\tau_k - \tau_{k-1}} E^* \left(\frac{\xi(s) - \xi(\tau_{k-1})}{s - \tau_{k-1}} \right) + \frac{\tau_k - s}{\tau_k - \tau_{k-1}} E^* \left(\frac{\xi(\tau_k) - \xi(s)}{\tau_k - s} \right).$$

It follows that

$$\begin{aligned} E^*V(\xi, 0, t) &\le \sum_{j=1}^{k-1} (\tau_j - \tau_{j-1}) E^* \left(\frac{\xi(\tau_j) - \xi(\tau_{j-1})}{\tau_j - \tau_{j-1}} \right) + (s - \tau_{k-1}) E^* \left(\frac{\xi(s) - \xi(\tau_{k-1})}{s - \tau_{k-1}} \right) \\ &\quad + (\tau_k - s) E^* \left(\frac{\xi(\tau_k) - \xi(s)}{\tau_k - s} \right) + \sum_{j=k+1}^{N} (\tau_j - \tau_{j-1}) E^* \left(\frac{\xi(\tau_j) - \xi(\tau_{j-1})}{\tau_j - \tau_{j-1}} \right) + \epsilon \\ &\le E^*V(\xi, 0, s) + E^*V(\xi, s, t) + \epsilon, \end{aligned}$$

where the sum from $j = 1, \ldots, k-1$ is only present when $k \ge 2$. If $s = \tau_{k-1}$ for some $1 \le k \le N$, we can arrive at the same inequality without introducing s as another point in the partition. Therefore, $E^*V(\xi, 0, t) \le E^*V(\xi, 0, s) + E^*V(\xi, s, t) + \epsilon$ in all cases. □

We now recall that since $\mathcal{P}^\perp$ is the dual of $C(\mathbb{S}^2)/\mathcal{P}$, $\xi : [0,\infty) \to \mathcal{P}^\perp$ is absolutely continuous if and only if ξ weak* differentiable almost everywhere, $\dot{\xi} \in L^1_{\text{loc}}([0,\infty); \mathcal{P}^\perp)$, and

$$\xi(t) - \xi(s) = \int_s^t \dot{\xi}(\tau)d\tau \tag{3.8}$$

for all $0 \le s < t$ (Remark 1.1.3 of [**1**]). We can use this characterization to show that if ξ is absolutely continuous then $E^*V(\xi, s, t)$ coincides with the integral that motivated it.

PROPOSITION 3.11. *Suppose $\xi : [0,\infty) \to \mathcal{P}^\perp$ is absolutely continuous. Then*

$$\int_s^t E^*(\dot{\xi}(\tau))d\tau = E^*V(\xi, s, t)$$

for all $0 \le s \le t$.

PROOF. Let $0 \le s < t$, and suppose $s = \tau_0 < \cdots < \tau_N = t$ is a partition of $[s,t]$. In view of (3.8) and Jensen's inequality

$$\begin{aligned}(\tau_k - \tau_{k-1})E^*\left(\frac{\xi(\tau_k) - \xi(\tau_{k-1})}{\tau_k - \tau_{k-1}}\right) &= (\tau_k - \tau_{k-1})E^*\left(\frac{\int_{\tau_{k-1}}^{\tau_k} \dot{\xi}(\tau)d\tau}{\tau_k - \tau_{k-1}}\right)\\ &\le \int_{\tau_{k-1}}^{\tau_k} E^*(\dot{\xi}(\tau))d\tau\end{aligned}$$

for $k = 1, \dots, N$. Therefore,

$$\sum_{k=1}^N (\tau_k - \tau_{k-1})E^*\left(\frac{\xi(\tau_k) - \xi(\tau_{k-1})}{\tau_k - \tau_{k-1}}\right) \le \int_s^t E^*(\dot{\xi}(\tau))d\tau.$$

It follows that

$$E^*V(\xi, s, t) \le \int_s^t E^*(\dot{\xi}(\tau))d\tau.$$

Now set

$$F(t) := E^*V(\xi, 0, t)$$

for $t \ge 0$. In view of the definition of E^* variation and (3.7),

$$\delta E^*\left(\frac{\xi(t+\delta) - \xi(t)}{\delta}\right) \le E^*V(\xi, t, t+\delta) = F(t+\delta) - F(t)$$

for $t \ge 0$ and $\delta > 0$. As F is monotone and ξ is weak* differentiable almost everywhere, we have

$$E^*(\dot{\xi}(\tau)) \le \dot{F}(t)$$

for almost every $t \ge 0$. Here we used that E^* is weak* lower-semicontinuous. That is,

$$\int_s^t E^*(\dot{\xi}(\tau))d\tau \le \int_s^t \dot{F}(t)d\tau \le F(t) - F(s) = E^*V(\xi, s, t)$$

for $0 \le s \le t$. □

COROLLARY 3.12. *Suppose $\tau > 0$ and $(\zeta^k)_{k\geq 0} \subset \mathcal{P}^\perp$. For $t \geq 0$, define*

$$\xi(t) := \zeta^{k-1} + \frac{t-(k-1)\tau}{\tau}(\zeta^k - \zeta^{k-1}), \quad (\tau-1)k \leq t \leq \tau k.$$

Then

$$E^*V(\xi, 0, N\tau) = \sum_{k=1}^{N} \tau E^*\left(\frac{\zeta^k - \zeta^{k-1}}{\tau}\right)$$

for each $N \in \mathbb{N}$.

PROOF. Observe that

$$\dot{\xi}(t) = \frac{\zeta^k - \zeta^{k-1}}{\tau}$$

for $t \in (\tau(k-1), \tau k)$, and

$$\int_{\tau(k-1)}^{\tau k} \|\dot{\xi}(t)\|_* dt = \|\zeta^k - \zeta^{k-1}\|_* < \infty$$

for each $k \in \mathbb{N}$. Furthermore, since ξ is piecewise linear, the fundamental theorem of calculus (3.8) holds. As a result, ξ is absolutely continuous. By the previous proposition,

$$E^*V(\xi, 0, N\tau) = \int_0^{N\tau} E^*(\dot{\xi}(\tau))d\tau = \sum_{k=1}^{N} \int_{\tau(k-1)}^{\tau k} E^*(\dot{\xi}(\tau))d\tau = \sum_{k=1}^{N} \tau E^*\left(\frac{\zeta^k - \zeta^{k-1}}{\tau}\right).$$

□

As $E^*(\zeta_1) = E^*(\zeta_2)$ whenever $\zeta_1 - \zeta_2 \in \mathcal{C}^\perp$, E^* can be viewed as a function on the quotient space $\mathcal{P}^\perp/\mathcal{C}^\perp$:

$$E^*(\boldsymbol{\zeta}) := \sup\{\langle \zeta, g\rangle - E(\mathbf{g}) : g \in \mathcal{C}\}$$

for $\boldsymbol{\zeta} \in \mathcal{P}^\perp/\mathcal{C}^\perp$. In particular, we may consider the E^* variation of paths $\boldsymbol{\xi} : [0,\infty) \to \mathcal{P}^\perp/\mathcal{C}^\perp$ and each of our results above applies to such paths. We finally note that this extension of E^* is lower-semicontinuous on $(\mathcal{P}^\perp/\mathcal{C}^\perp, \chi^*_{\mathcal{C}})$.

LEMMA 3.13. *Suppose $(\boldsymbol{\xi}^k)_{k\in\mathbb{N}}$ is a sequence in $\mathcal{P}^\perp/\mathcal{C}^\perp$ with*

$$\lim_{k\to\infty} \chi^*_{\mathcal{C}}(\boldsymbol{\xi}^k - \boldsymbol{\xi}) = 0$$

Then

$$\liminf_{k\to\infty} E^*(\boldsymbol{\xi}^k) \geq E^*(\boldsymbol{\xi}).$$

PROOF. Let $g \in \mathcal{C}$. Observe that

$$|\langle \xi^k, g\rangle - \langle \xi, g\rangle| = |\langle \xi^k - \xi, g\rangle| \leq \chi^*_{\mathcal{C}}(\boldsymbol{\xi}^k - \boldsymbol{\xi}).$$

Therefore, $\langle \xi^k, g\rangle \to \langle \xi, g\rangle$ as $k \to \infty$. This implies

$$\liminf_{k\to\infty} E^*(\boldsymbol{\xi}^k) \geq \liminf_{k\to\infty}(\langle \xi^k, g\rangle - E(\mathbf{g})) = \langle \xi, g\rangle - E(\mathbf{g}).$$

We conclude upon taking the supremum of $g \in \mathcal{C}$. □

4. Existence theorem

We will now analyze solutions $\xi : [0,\infty) \to \mathcal{P}^{\perp}$ of the equation

$$\partial E^*(\dot{\xi}(t)) + \mathcal{J}^*(\xi(t)) \ni 0 \quad \text{a.e. } t \geq 0. \tag{4.1}$$

These types of flows have been considered in models for diverse phenomena such phase transitions [**9, 12, 53**], fracture mechanics [**22, 32, 39**], and hysteresis effects [**40, 52**]. The existence of solutions to doubly nonlinear evolutions in Hilbert spaces [**5, 7, 16**] and reflexive Banach spaces [**17**] for given initial conditions were established many years ago. The reflexivity requirement can be relaxed using the notion of "curves of maximal slope" as described in Chapter 1 of [**1**].

All of these results require that the corresponding E^* is coercive or superlinear with respect to the norm; this would ensure that solutions are absolutely continuous. We cannot use any of these results directly as our E^* functional is not coercive. A similar problem has been encountered in the theory of rate-independent doubly nonlinear evolutions [**18, 31, 35, 40–44, 48**]. Here the corresponding E^* function is a norm and solutions are typically considered as mappings of bounded variation measured in this norm. We will adapt this approach to our problem using the E^* variation notion discussed in the previous section.

We will define a type of weak solution of (4.1) below. To avoid confusion, we'll say that ξ is a *classical solution* of (4.1) if it satisfies the equation as described in the introduction. That is, ξ is a classical solution if it is absolutely continuous and if there is a measurable $\mathbf{g} : [0,\infty) \to C(\mathbb{S}^2)/\mathcal{P}$ such that

$$\mathbf{g}(t) \in \mathcal{J}^*(\xi(t)) \cap (-\partial E^*(\dot{\xi}(t))) \tag{4.2}$$

for almost every $t \geq 0$. Here $\dot{\xi}(t)$ is the weak* derivative of ξ at a time $t \geq 0$.

DEFINITION 4.1. A measurable mapping

$$\xi : [0,\infty) \to \mathcal{P}^{\perp}$$

with locally finite E^* variation is a *weak solution* of the doubly nonlinear evolution (4.1) provided there is a measurable

$$\mathbf{g} : [0,\infty) \to C(\mathbb{S}^2)/\mathcal{P}$$

such that

$$\frac{1}{2}\|\xi(s)\|_*^2 - \frac{1}{2}\|\xi(t)\|_*^2 = E^*V(\xi, s, t) + \int_s^t E(\mathbf{g}(\tau))d\tau$$

for almost every $0 \leq s \leq t$ and

$$\mathbf{g}(t) \in \partial\mathcal{J}^*(\xi(t))$$

for almost every $t \geq 0$.

For any weak solution ξ, $\|\xi(t)\|_*$ is a nonincreasing function of t outside of a null set. We will say such functions are *essentially nonincreasing*. In the appendix, we will recall that an essentially nonincreasing function is simply a nonincreasing function modified on a null set (Lemma B.1). Also note that by the duality formula (1.1)

$$[0,\infty) \ni t \mapsto \|\mathbf{g}(t)\|$$

is essentially nonincreasing, as well.

Let us check that the notion of weak solutions extends the classical notion of solution.

LEMMA 4.2. *Assume* $\xi : [0,\infty) \to \mathcal{P}^\perp$ *is a classical solution of* (4.1). *Then* ξ *is a weak solution of* (4.1).

PROOF. Since ξ is absolutely continuous, it is measurable. We may also select a measurable $\mathbf{g} : [0,\infty) \to C(\mathbb{S}^2)/\mathcal{P}$ such that (4.2) holds almost everywhere on $[0,\infty)$. Note that for almost every $t > 0$,

$$\frac{1}{2}\|\xi(t+\tau)\|_*^2 \geq \frac{1}{2}\|\xi(t)\|_*^2 + \langle \xi(t+\tau) - \xi(t), g(t)\rangle$$

for $\tau \in \mathbb{R}$ sufficiently small. It follows that

$$\lim_{\tau\to 0^+} \frac{\|\xi(t+\tau)\|_*^2 - \|\xi(t)\|_*^2}{2\tau} \geq \langle \dot{\xi}(t), g(t)\rangle \geq \lim_{\tau\to 0^-} \frac{\|\xi(t+\tau)\|_*^2 - \|\xi(t)\|_*^2}{2\tau}$$

for almost every $t > 0$. As $[0,\infty) \ni t \to \|\xi(t)\|_*$ is absolutely continuous,

$$\frac{d}{dt}\frac{1}{2}\|\xi(t)\|_*^2 = \langle \dot{\xi}(t), g(t)\rangle$$

almost everywhere in $[0,\infty)$.

Recall that $-\dot{\xi}(t) \in \partial E(\mathbf{g}(t)) \neq \emptyset$ for almost every $t \geq 0$. As a result, $g(t) \in \mathcal{C}$ and

$$\frac{d}{dt}\frac{1}{2}\|\xi(t)\|_*^2 = -\langle \dot{\xi}(t), -g(t)\rangle = -[E^*(\dot{\xi}(t)) + E(\mathbf{g}(t))]$$

for almost every $t \geq 0$. Integrating this identity from s to t with $s \leq t$ gives

$$\begin{aligned}\frac{1}{2}\|\xi(s)\|_*^2 - \frac{1}{2}\|\xi(t)\|_*^2 &= \int_s^t E^*(\dot{\xi}(\tau))d\tau + \int_s^t E(\mathbf{g}(\tau))d\tau \\ &= E^*V(\xi,s,t) + \int_s^t E(\mathbf{g}(\tau))d\tau.\end{aligned}$$

Here we used Proposition 3.11. We conclude that that ξ is indeed a weak solution. □

The main theorem of this paper as follows.

THEOREM 4.3. *Assume* $g^0 \in \mathcal{C}$ *and that* $\xi^0 \in \mathcal{J}(\mathbf{g}^0)$. *There is a weak solution* ξ *with finite* E^* *variation which satisfies the initial condition*

$$\xi(0)|_{\mathcal{C}} = \xi^0|_{\mathcal{C}}. \tag{4.3}$$

Moreover, there is a corresponding $\mathbf{g} : [0,\infty) \to C(\mathbb{S}^2)/\mathcal{P}$ *as described in Definition* 4.1 *for which* $\mathbf{g} \in L^2([0,\infty); \mathcal{V}/\mathcal{P})$ *and*

$$[0,\infty) \ni t \mapsto E(\mathbf{g}(t)) \quad \textit{is essentially nonincreasing.}$$

As a corollary to the theorem above, we have that $tE(\mathbf{g}(t))$ and $\|\mathbf{g}(t)\|$ both converge to zero along sequences to times t tending to infinity outside of a null set.

COROLLARY 4.4. *There is a null set* $N \subset [0,\infty)$ *such that*

$$\lim_{\substack{t\to\infty \\ t\notin N}} tE(\mathbf{g}(t)) = 0 \quad \textit{and} \quad \lim_{\substack{t\to\infty \\ t\notin N}} \|\mathbf{g}(t)\| = 0. \tag{4.4}$$

PROOF. As $E \circ \mathbf{g} : [0,\infty) \to [0,\infty)$ is integrable and essentially nonincreasing, Proposition (B.2) of the appendix implies there is a null set N for which

$$\lim_{\substack{t\to\infty \\ t\notin N}} tE(\mathbf{g}(t)) = 0. \tag{4.5}$$

Without any loss of generality, we may assume $g(t) \in \mathcal{C}$ for all $t \notin N$ since this occurs for almost every time $t \geq 0$. Let $(t_k)_{k\in\mathbb{N}} \subset [0,\infty)\setminus N$ be a sequence increasing to ∞. By Lemma 3.1, $(\mathbf{g}(t_k))_{k\in\mathbb{N}}$ has a convergent subsequence $(\mathbf{g}(t_{k_j}))_{j\in\mathbb{N}}$. Furthermore, there is a sequence $(g^j)_{j\in\mathbb{N}} \subset C^1(\mathbb{S}^2)$ with $g^j \in \mathbf{g}(t_{k_j})$ and which converges in $C^1(\mathbb{S}^2)$ to some limit function g^∞. By (4.5),

$$E(\mathbf{g}(t_{k_j})) = \int_{\mathbb{S}^2} \left(\frac{1}{2}|\nabla g^j|^2 - (g^j)^2\right) d\sigma \leq \frac{1}{t_{k_j}}$$

for all large enough j. Thus,

$$\int_{\mathbb{S}^2} \left(\frac{1}{2}|\nabla g^\infty|^2 - (g^\infty)^2\right) d\sigma = \lim_{j\to\infty} \left(\int_{\mathbb{S}^2} \frac{1}{2}|\nabla g^j|^2 - (g^j)^2\right) d\sigma = 0.$$

This implies $g^\infty(u) = b \cdot u$ for some $b \in \mathbb{R}^3$. It then follows that

$$\lim_{j\to\infty} \|\mathbf{g}(t_{k_j})\| \leq \lim_{j\to\infty} \max_{|u|=1} |g^j(u) - b \cdot u| = 0.$$

Since the sequence $(t_k)_{k\in\mathbb{N}} \subset [0,\infty)\setminus N$ was arbitrary, we conclude (4.4). □

We may interpret these results geometrically. Choose a null set $N \subset [0,\infty)$ such that for all $t \in N^c$, $g(t) \in \mathcal{C}$. For each $t \in N^c$, $g(t) + 1/2$ is the $\mathbb{S}^2$ restriction of the support function of a constant width body K_t. Then Theorem 4.3, the limits (4.4), and equation (2.4) imply the following assertion on the family $(K_t)_{t\in N^c}$.

COROLLARY 4.5. (i) $R(K_t) = \frac{1}{2} + \|\mathbf{g}(t)\|$ *is essentially nonincreasing, and*

$$\lim_{\substack{t\to\infty\\ t\notin N}} R(K_t) = \frac{1}{2}.$$

(ii) $V(K_t) = \frac{\pi}{6} - \frac{1}{2}E(\mathbf{g}(t))$ *is essentially nondecreasing, and*

$$\lim_{\substack{t\to\infty\\ t\notin N}} t\left(\frac{\pi}{6} - V(K_t)\right) = 0.$$

(iii) *Suppose* $a(t) \in \mathbb{R}^3$ *is the center of the circumball of* K_t *for each* $t \in N^c$. *Then*

$$\lim_{\substack{t\to\infty\\ t\notin N}} d_{\mathcal{H}}(K_t - a(t), B_{1/2}(0)) = 0$$

We will design a weak solution as asserted in Theorem 4.3 as follows. First, we will show how to solve a discrete version of (4.1) which depends on a parameter $\tau > 0$. Next, we will show how to use these solutions to form a family of approximate solutions of (4.1) indexed by τ. Then we will derive various inequalities satisfied by this family of approximate solutions which are independent of τ. Finally, we will explain how to extract a sequence of τ tending to 0 for which the approximate solutions converge to a weak solution.

4.1. Implicit time scheme. For the remainder of this section we will suppose the hypotheses of Theorem 4.3. That is, we will assume $g^0 \in \mathcal{C}$ and that $\xi^0 \in \mathcal{J}(\mathbf{g}^0)$. In order to prove this theorem, we will use the following implicit time scheme: fix $\tau > 0$ and find a sequence $(\xi^k)_{k\in\mathbb{N}} \subset \mathcal{P}^\perp$ such that

$$\partial E^*\left(\frac{\xi^k - \xi^{k-1}}{\tau}\right) + \mathcal{J}^*(\xi^k) \ni 0 \tag{4.6}$$

for $k \in \mathbb{N}$. Let us verify that there is indeed a solution sequence.

LEMMA 4.6. *There is a solution sequence* $(\xi^k)_{k\in\mathbb{N}}$ *of* (4.6) *which satisfies*

$$\|\xi^k\|_* = \|\boldsymbol{\xi}^k\|_* \tag{4.7}$$

for each $k \in \mathbb{N}$.

REMARK 4.7. Here $\boldsymbol{\xi}^k = \xi^k + \mathcal{C}^\perp$ is the equivalence class defined in subsection 3.3.

PROOF. We will proceed by induction. Once $\xi^0, \dots, \xi^{k-1} \in \mathcal{P}^\perp$ are determined, we can minimize

$$\mathcal{P}^\perp \ni \xi \mapsto \tau E^*\left(\frac{\xi - \xi^{k-1}}{\tau}\right) + \frac{1}{2}\|\xi\|_*^2.$$

Starting with any minimizing sequence, we can employ Alaoglu's theorem and the weak* lower-semicontinuity of both E^* and the norm on $\mathcal{P}^\perp$ to conclude the existence of a minimizer $\xi^k \in \mathcal{P}^\perp$. Therefore, there is a solution sequence $(\xi^k)_{k\in\mathbb{N}}$ of (4.6).

We may select

$$\mathbf{g}^k \in \mathcal{J}^*(\xi^k) \cap \left(-\partial E^*\left(\frac{\xi^k - \xi^{k-1}}{\tau}\right)\right)$$

for $k \in \mathbb{N}$. Note that since

$$-\frac{\xi^k - \xi^{k-1}}{\tau} \in \partial E(\mathbf{g}^k) \neq \emptyset, \tag{4.8}$$

$g^k \in \mathcal{C}$ for each $k \in \mathbb{N}$, as well. Therefore, if $\eta \in \mathcal{C}^\perp$, then

$$\frac{1}{2}\|\xi^k + \eta\|_*^2 \geq \frac{1}{2}\|\xi^k\|_*^2 + \langle \eta, g^k \rangle = \frac{1}{2}\|\xi^k\|_*^2.$$

We conclude (4.7). □

As we saw for the classical solutions of (4.1), the sequences $(\xi^k)_{k\in\mathbb{N}} \subset \mathcal{P}^\perp$ and $(\mathbf{g}^k)_{k\in\mathbb{N}} \subset C(\mathbb{S}^2)/\mathcal{P}$ have two important monotonicity properties.

LEMMA 4.8. *For each* $0 \leq j < k$,

$$\frac{1}{2}\|\xi^j\|_*^2 \geq \frac{1}{2}\|\xi^k\|_*^2 + \sum_{\ell=j+1}^{k} \tau\left[E^*\left(\frac{\xi^\ell - \xi^{\ell-1}}{\tau}\right) + E(\mathbf{g}^\ell)\right] \tag{4.9}$$

and

$$E(\mathbf{g}^j) \geq E(\mathbf{g}^k) + \sum_{\ell=j+1}^{k} \left\langle \frac{\xi^k - \xi^{k-1}}{\tau}, g^k - g^{k-1} \right\rangle. \tag{4.10}$$

PROOF. First note

$$\begin{aligned}
\frac{1}{2}\|\xi^{\ell-1}\|_*^2 &\geq \frac{1}{2}\|\xi^\ell\|_*^2 + \langle \xi^{\ell-1} - \xi^\ell, g^\ell \rangle \\
&= \frac{1}{2}\|\xi^\ell\|_*^2 + \tau\left\langle \frac{\xi^\ell - \xi^{\ell-1}}{\tau}, -g^\ell \right\rangle \\
&= \frac{1}{2}\|\xi^\ell\|_*^2 + \tau\left[E^*\left(\frac{\xi^\ell - \xi^{\ell-1}}{\tau}\right) + E(-\mathbf{g}^\ell)\right] \\
&= \frac{1}{2}\|\xi^\ell\|_*^2 + \tau\left[E^*\left(\frac{\xi^\ell - \xi^{\ell-1}}{\tau}\right) + E(\mathbf{g}^\ell)\right].
\end{aligned}$$

Then (4.9) follows from summing from $\ell = j+1$ to k.

Next observe that in view of (4.8),

$$\begin{aligned} E(\mathbf{g}^{\ell-1}) &\geq E(\mathbf{g}^{\ell}) + \left\langle -\frac{\xi^{\ell} - \xi^{\ell-1}}{\tau}, g^{\ell-1} - g^{\ell} \right\rangle \\ &= E(\mathbf{g}^{\ell}) + \left\langle \frac{\xi^{\ell} - \xi^{\ell-1}}{\tau}, g^{\ell} - g^{\ell-1} \right\rangle. \end{aligned}$$

Inequality (4.10) results from summing from $\ell = j+1$ to k. □

REMARK 4.9. Since

$$\|\xi^k\|_* = \|\mathbf{g}^k\|,$$

it follows that $\left(\|\mathbf{g}^k\|\right)_{k\in\mathbb{N}}$ is nonincreasing.

In designing a solution, it will help to identify some key variables. We will denote

$$\xi_\tau(t) := \xi^{k-1} + \left(\frac{t-(k-1)\tau}{\tau}\right)\left(\xi^k - \xi^{k-1}\right)$$

for $t \in [(k-1)\tau, k\tau]$ and

$$\zeta_\tau(t) := \begin{cases} \xi^0, & t = 0 \\ \xi^k, & t \in ((k-1)\tau, k\tau] \end{cases}$$

for $t \geq 0$. In addition, we will consider

$$\mathbf{g}_\tau(t) := \begin{cases} \mathbf{g}^0, & t = 0 \\ \mathbf{g}^k, & t \in ((k-1)\tau, k\tau] \end{cases}$$

for $t \geq 0$. In terms of these variables, we note

$$\partial E^*\left(\dot{\xi}_\tau(t)\right) + \mathcal{J}^*(\zeta_\tau(t)) \ni 0$$

and

$$\mathbf{g}_\tau(t) \in \mathcal{J}^*(\zeta_\tau(t)) \cap \left(-\partial E^*\left(\dot{\xi}_\tau(t)\right)\right)$$

for $t \neq k\tau$.

4.2. Various bounds. We will now derive various bounds on the variables we defined. We will see that the following proposition essentially follows from the monotonicity formula (4.9) and (4.10).

PROPOSITION 4.10. *For $\tau > 0$ and $t \geq 0$, the following inequalities hold.*

$$\|\zeta_\tau(t)\|_* \leq \|\xi^0\|_* \tag{4.11}$$

$$\|\xi_\tau(t)\|_* \leq \|\xi^0\|_* \tag{4.12}$$

$$E^*V(\xi_\tau, 0, \infty) \leq \frac{1}{2}\|\xi^0\|_*^2 \tag{4.13}$$

$$\int_0^\infty E(\mathbf{g}_\tau(t))dt \leq \frac{1}{2}\|\xi^0\|_*^2 \tag{4.14}$$

$$E(\mathbf{g}_\tau(t)) \leq E(\mathbf{g}^0) \tag{4.15}$$

PROOF. In view of (4.9), $\|\xi^k\|_* \leq \|\xi^{k-1}\|_* \leq \|\xi^0\|_*$ for $k \in \mathbb{N}$. It is now immediate that (4.11) holds. Likewise for $t \in [(k-1)\tau, k\tau]$,

$$\begin{aligned}\|\xi_\tau(t)\|_* &= \left\|\left(1 - \frac{t-(k-1)\tau}{\tau}\right)\xi^{k-1} + \left(\frac{t-(k-1)\tau}{\tau}\right)\xi^k\right\|_* \\ &\leq \left(1 - \frac{t-(k-1)\tau}{\tau}\right)\|\xi^{k-1}\|_* + \left(\frac{t-(k-1)\tau}{\tau}\right)\|\xi^k\|_* \\ &\leq \|\xi^{k-1}\|_* \\ &\leq \|\xi^0\|_*.\end{aligned}$$

We conclude (4.12).

In view of Corollary 3.12 and (4.9),

$$E^*V(\xi_\tau, 0, N\tau) = \sum_{k=1}^{N} \tau E^*\left(\frac{\zeta^k - \zeta^{k-1}}{\tau}\right) \leq \frac{1}{2}\|\xi^0\|_*^2$$

for each $N \in \mathbb{N}$. Sending $N \to \infty$ gives (4.13). Likewise, we have

$$\int_0^{N\tau} E(\mathbf{g}_\tau(t))dt = \sum_{k=1}^{N} E(\mathbf{g}^k)\tau \leq \frac{1}{2}\|\xi^0\|_*^2$$

for every $N \in \mathbb{N}$. And sending $N \to \infty$ gives (4.14).

By (4.10), $E(\mathbf{g}^k) \leq E(\mathbf{g}^0)$ for $k \in \mathbb{N}$. Therefore, (4.15) holds. □

It is natural to anticipate that ξ_τ and ζ_τ are close. We can measure their closeness via their respective equivalence classes in $\mathcal{P}^\perp/\mathcal{C}^\perp$.

PROPOSITION 4.11. *For each $T \geq 0$,*

$$\int_0^T \chi_{\mathcal{C}}^*(\boldsymbol{\xi}_\tau(t) - \boldsymbol{\zeta}_\tau(t))dt \leq \frac{\tau}{2}\left[\frac{1}{2}\|\xi^0\|_*^2 + \frac{\pi}{3}(T+\tau)\right].$$

PROOF. Let $g \in \mathcal{C}$ and recall that $E(\mathbf{g}) \leq \pi/3$ by (1.3). Note that for $t \in ((k-1)\tau, k\tau]$

$$\begin{aligned}\langle \xi_\tau(t) - \zeta_\tau(t), g\rangle &= \tau\left\langle \frac{\xi^k - \xi^{k-1}}{\tau}, g\right\rangle\left(\frac{t-k\tau}{\tau}\right) \\ &\leq \tau\left[E^*\left(\frac{\xi^k - \xi^{k-1}}{\tau}\right) + E(\mathbf{g})\right]\left(\frac{k\tau - t}{\tau}\right) \\ &\leq \tau\left[E^*\left(\frac{\xi^k - \xi^{k-1}}{\tau}\right) + \frac{\pi}{3}\right]\left(\frac{k\tau - t}{\tau}\right).\end{aligned}$$

As a result,

$$\chi_{\mathcal{C}}^*(\boldsymbol{\xi}_\tau(t) - \boldsymbol{\zeta}_\tau(t)) \leq \tau\left[E^*\left(\frac{\xi^k - \xi^{k-1}}{\tau}\right) + \frac{\pi}{3}\right]\left(\frac{k\tau - t}{\tau}\right).$$

Integrating over $[(k-1)\tau, k\tau]$ gives

$$\int_{(k-1)\tau}^{k\tau} \chi_{\mathcal{C}}^*(\boldsymbol{\xi}_\tau(t) - \boldsymbol{\zeta}_\tau(t))dt \leq \frac{1}{2}\tau^2\left[E^*\left(\frac{\xi^k - \xi^{k-1}}{\tau}\right) + \frac{\pi}{3}\right]$$

for each $k \in \mathbb{N}$.

Now let $T > 0$ and choose $N \in \mathbb{N}$ so that

$$N - 1 < \frac{T}{\tau} \leq N.$$

In view of (4.9),

$$\begin{aligned}
\int_0^T \chi_C^*(\boldsymbol{\xi}_\tau(t) - \boldsymbol{\zeta}_\tau(t))dt &\leq \int_0^{N\tau} \chi_C^*(\boldsymbol{\xi}_\tau(t) - \boldsymbol{\zeta}_\tau(t))dt \\
&= \sum_{k=1}^N \int_{(k-1)\tau}^{k\tau} \chi_C^*(\boldsymbol{\xi}_\tau(t) - \boldsymbol{\zeta}_\tau(t))dt \\
&\leq \sum_{k=1}^N \frac{1}{2}\tau^2 \left[E^*\left(\frac{\xi^k - \xi^{k-1}}{\tau}\right) + \frac{\pi}{3}\right] \\
&\leq \frac{\tau}{2}\left[\frac{1}{2}\|\xi^0\|_*^2 + (N\tau)\frac{\pi}{3}\right] \\
&\leq \frac{\tau}{2}\left[\frac{1}{2}\|\xi^0\|_*^2 + (T + \tau)\frac{\pi}{3}\right].
\end{aligned}$$

□

A simple continuity estimate that will prove to be very useful to us is as follows.

PROPOSITION 4.12. *For* $0 \leq s < t$,

$$\chi_C^*(\boldsymbol{\xi}_\tau(t) - \boldsymbol{\xi}_\tau(s)) \leq E^*V(\xi_\tau, s, t) + \frac{\pi}{3}(t - s).$$

PROOF. Let $g \in \mathcal{C}$ and observe

$$\begin{aligned}
\langle \xi_\tau(t) - \xi_\tau(s), g\rangle &= (t - s)\left\langle \frac{\xi_\tau(t) - \xi_\tau(s)}{t - s}, g\right\rangle \\
&\leq (t - s)E^*V\left(\frac{\xi_\tau(t) - \xi_\tau(s)}{t - s}\right) + (t - s)E(\mathbf{g}) \\
&\leq E^*V(\xi_\tau, s, t) + \frac{\pi}{3}(t - s).
\end{aligned}$$

We conclude by taking the supremum over $g \in \mathcal{C}$. □

4.3. Compactness. In this subsection, we will establish various assertions involving the convergence of a given quantity along a subsequence of $\tau_j \to 0^+$. Since there will be only finitely many of these types of statements, we will not alter the subsequence for each additional limiting assertion which may only guarantee that a limit holds upon passing to a further subsequence.

In view of (4.9) and (4.11),

$$\|\zeta_\tau(t)\|_* \leq \|\zeta_\tau(s)\|_* \leq \|\xi^0\|_*$$

for $0 \leq s \leq t$. Likewise (4.13) implies

$$E^*V(\xi_\tau, 0, s) \leq E^*V(\xi_\tau, 0, t) \leq \frac{1}{2}\|\xi^0\|_*^2$$

for $0 \le s \le t$. By Helly's selection theorem, there is a sequence $(\tau_j)_{j\in\mathbb{N}}$ decreasing to 0 for which the limits

$$
\begin{cases} A(t) := \lim_{j\to\infty} \|\zeta_{\tau_j}(t)\|_* \\ \\ B(t) := \lim_{j\to\infty} E^*V(\xi_{\tau_j}, 0, t) \end{cases} \tag{4.16}
$$

exist for all $t \ge 0$.

By (4.14), $(\mathbf{g}_\tau)_{\tau>0} \subset L^2([0,\infty); \mathcal{V}/\mathcal{P})$ is bounded. Therefore, passing to a subsequence if necessary,

$$
(\mathbf{g}_{\tau_j})_{j\in\mathbb{N}} \text{ converges weakly to some } \mathbf{g} \in L^2([0,\infty); \mathcal{V}/\mathcal{P}).
$$

Recall that $g(t) \in \mathcal{C}$ for almost all $t \ge 0$ by Lemma 3.4. In view of Lemma 3.5, we may also assume

$$
\begin{cases} g(t) \in \mathcal{C} \text{ for all } t \ge 0 \text{ and} \\ \mathbf{g} : [0,\infty) \to C(\mathbb{S}^2)/\mathcal{P} \text{ is measurable} \end{cases}
$$

without any loss of generality. We shall do so going forward.

Our most important convergence assertion is as follows.

PROPOSITION 4.13. *There is $\boldsymbol{\xi} : [0,\infty) \to \mathcal{P}^\perp/\mathcal{C}^\perp$ with finite E^* variation for which we can pass to a subsequence if necessary to obtain*

$$
\lim_{j\to\infty} \chi^*_{\mathcal{C}}(\boldsymbol{\xi}_{\tau_j}(t) - \boldsymbol{\xi}(t)) = 0 \tag{4.17}
$$

and

$$
\lim_{j\to\infty} \int_0^t \chi^*_{\mathcal{C}}(\boldsymbol{\zeta}_{\tau_j}(s) - \boldsymbol{\xi}(s))ds = 0 \tag{4.18}
$$

for all $t \ge 0$, and

$$
\lim_{s\to t} \chi^*_{\mathcal{C}}(\boldsymbol{\xi}(t) - \boldsymbol{\xi}(s)) = 0 \tag{4.19}
$$

for all but countably many $t \ge 0$.

PROOF. By (4.12),

$$
\|\boldsymbol{\xi}_\tau(t)\|_* \le \|\xi_\tau(t)\|_* \le \|\xi^0\|_*;
$$

that is,

$$
\boldsymbol{\xi}_\tau(t) \in \{\boldsymbol{\zeta} \in \mathcal{P}^\perp/\mathcal{C}^\perp : \|\boldsymbol{\zeta}\|_* \le \|\xi^0\|_*\} := \mathcal{B}_{\|\xi^0\|_*}
$$

for all $t \ge 0$ and $\tau > 0$. By Proposition 3.7, $(\mathcal{B}_{\|\xi^0\|_*}, \chi^*_{\mathcal{C}})$ is a compact metric space.

We now appeal to Proposition 4.12 which asserts

$$
\begin{aligned}
\chi^*_{\mathcal{C}}(\boldsymbol{\xi}_\tau(t) - \boldsymbol{\xi}_\tau(s)) &\le E^*V(\xi_\tau, s, t) + \frac{\pi}{3}(t-s) \\
&= E^*V(\xi_\tau, 0, t) - E^*V(\xi_\tau, 0, s) + \frac{\pi}{3}(t-s)
\end{aligned}
$$

for $0 \le s \le t$. For the last equality, we used Lemma 3.10. In view of (4.16),

$$
\limsup_{j\to\infty} \chi^*_{\mathcal{C}}(\boldsymbol{\xi}_{\tau_j}(t) - \boldsymbol{\xi}_{\tau_j}(s)) \le B(t) - B(s) + \frac{\pi}{3}(t-s)
$$

for all $s \le t$. Since $B : [0,\infty) \to [0,\infty)$ is nondecreasing, it is continuous for all but countably many $t \ge 0$. By an abstract version of the Arzelà-Ascoli theorem (Proposition 3.3.1 in [**1**]), there is a mapping $\boldsymbol{\xi} : [0,\infty) \to \mathcal{B}_{\|\xi^0\|_*}$ satisfying (4.19) for all but countably many $t \geq 0$. Moreover, along an appropriate subsequence of $(\boldsymbol{\xi}_{\tau_j})_{j\in\mathbb{N}}$, (4.17) holds for every $t \geq 0$.

By Lemma 3.13, E^* is lower-semicontinuous with respect to convergence in the metric induced by $\chi^*_{\mathcal{C}}$. In view of the pointwise convergence (4.17),

$$E^*V(\boldsymbol{\xi}, 0, t) \leq \varliminf_{j\to\infty} E^*V(\boldsymbol{\xi}_{\tau_j}, 0, t) \leq \frac{1}{2}\|\xi^0\|_*^2$$

for $t \geq 0$. Therefore, $\boldsymbol{\xi}$ has finite E^* variation.

Recall inequality (3.5). This implies

$$\chi^*_{\mathcal{C}}(\boldsymbol{\xi}_{\tau_j}(s) - \boldsymbol{\xi}(s)) \leq c\|\boldsymbol{\xi}_{\tau_j}(s) - \boldsymbol{\xi}(s)\|_* \leq c(\|\boldsymbol{\xi}_{\tau_j}(s)\| + \|\boldsymbol{\xi}(s)\|_*) \leq 2c\|\xi^0\|_*$$

for each $s \geq 0$, where $c = \sqrt{\frac{3}{8} - \frac{1}{2}}$. As a result, we can use dominated convergence to conclude

$$\lim_{j\to\infty} \int_0^t \chi^*_{\mathcal{C}}(\boldsymbol{\xi}_{\tau_j}(s) - \boldsymbol{\xi}(s))ds = 0 \tag{4.20}$$

for each $t \geq 0$.

Let us also recall Proposition 4.11, which gives

$$\int_0^t \chi^*_{\mathcal{C}}(\boldsymbol{\xi}_{\tau_j}(s) - \boldsymbol{\zeta}_{\tau_j}(s))ds \leq \frac{\tau_j}{2}\left[\frac{1}{2}\|\xi^0\|_*^2 + \frac{\pi}{3}(t+\tau_j)\right].$$

Employing the triangle inequality,

$$\int_0^t \chi^*_{\mathcal{C}}(\boldsymbol{\zeta}_{\tau_j}(s) - \boldsymbol{\xi}(s))ds \leq \frac{\tau_j}{2}\left[\frac{1}{2}\|\xi^0\|_*^2 + \frac{\pi}{3}(t+\tau_j)\right] + \int_0^t \chi^*_{\mathcal{C}}(\boldsymbol{\xi}_{\tau_j}(s) - \boldsymbol{\xi}(s))ds.$$

We can now use (4.20) and send $j \to \infty$ to deduce (4.18). □

COROLLARY 4.14. *There is a measurable $\xi : [0,\infty) \to \mathcal{P}^\perp$ with finite E^* variation such that $\xi(t) \in \boldsymbol{\xi}(t)$ and*

$$\|\boldsymbol{\xi}(t)\|_* = \|\xi(t)\|_*$$

for all $t \geq 0$.

PROOF. We first claim that given any $\mu \in \mathcal{P}^\perp$, the function

$$f(t) := \inf\{\|\mu - \zeta\|_* : \zeta \in \boldsymbol{\xi}(t)\} \quad (t \geq 0)$$

is Lebesgue measurable. Note that we may rewrite this function as

$$f(t) = \inf\{\|\xi(t) - \mu + \eta\|_* : \eta \in \mathcal{C}^\perp\} = \|\boldsymbol{\xi}(t) - \boldsymbol{\mu}\|_*.$$

By Corollary 3.8, the norm on $\mathcal{P}^\perp/\mathcal{C}^\perp$ is lower-semicontinuous with convergence with respect to the metric induced by $\chi^*_{\mathcal{C}}$. Combined with the continuity of $\boldsymbol{\xi}$ (4.19), f is lower-semicontinuous for all but countably many times $t \geq 0$.

By Lemma C.1 in the appendix, f is necessarily Lebesgue measurable. Since $\boldsymbol{\xi}(t) \subset \mathcal{P}^\perp$ is closed and nonempty for each $t \geq 0$, it follows that

$$\boldsymbol{\xi} : [0,\infty) \rightsquigarrow \mathcal{P}$$

is a measurable, set-valued mapping (Theorem 8.1.4 of [**6**]). As a result, the mapping

$$\boldsymbol{\zeta} : [0,\infty) \rightsquigarrow \mathcal{P}$$

defined by

$$\boldsymbol{\zeta}(t) := \{\zeta \in \boldsymbol{\xi}(t) : \|\zeta\|_* = \|\boldsymbol{\xi}(t)\|_*\} \quad (t \geq 0)$$

is measurable (Theorem 8.2.11 of [**6**]). Consequently, there is a measurable $\xi : [0,\infty) \to \mathcal{P}^{\perp}$ for which

$$\xi(t) \in \boldsymbol{\zeta}(t)$$

and

$$\|\xi(t)\|_* = \|\boldsymbol{\xi}(t)\|_*$$

for all $t \geq 0$. And since $E^*(\boldsymbol{\xi}, 0, t) = E^*(\xi, 0, t)$ for all $t \geq 0$, ξ has finite E^* variation. □

REMARK 4.15. We also note that as $\xi(0) \in \boldsymbol{\xi}(0) = \xi^0 + \mathcal{C}^{\perp}$, ξ satisfies the initial value condition (4.3).

COROLLARY 4.16. *For almost every $t \geq 0$,*

$$\lim_{j\to\infty} \|\zeta_{\tau_j}(t)\|_* = \|\xi(t)\|_*.$$

PROOF. We have already noted that the limit

$$A(t) = \lim_{j\to\infty} \|\zeta_{\tau_j}(t)\|_*$$

exists for all $t \geq 0$. In view of (4.18), we may assume, after passing to a further subsequence if necessary, that

$$\lim_{j\to\infty} \chi^*_{\mathcal{C}}(\boldsymbol{\zeta}_{\tau_j}(t) - \boldsymbol{\xi}(t)) = 0$$

for almost every $t \geq 0$. Recall that

$$\|\zeta_{\tau_j}(t)\|_* = \|\boldsymbol{\zeta}_{\tau_j}(t)\|_*$$

by (4.7) for all but countable many $t \geq 0$. Therefore,

$$A(t) = \lim_{j\to\infty} \|\boldsymbol{\zeta}_{\tau_j}(t)\|_* \geq \|\boldsymbol{\xi}(t)\|_* = \|\xi(t)\|_*$$

for almost every $t \geq 0$. Here used Corollary 3.8.

Since $\mathbf{g}_\tau(t) \in \mathcal{J}^*(\xi_\tau(t))$ for almost all $t \geq 0$, we also have

$$\begin{aligned}\frac{1}{2}\|\xi(t)\|_*^2 &\geq \frac{1}{2}\|\zeta_{\tau_j}(t)\|_*^2 + \langle \xi(t) - \zeta_{\tau_j}(t), g_{\tau_j}(t)\rangle \\ &\geq \frac{1}{2}\|\zeta_{\tau_j}(t)\|_*^2 - \chi^*_{\mathcal{C}}(\boldsymbol{\xi}(t) - \boldsymbol{\zeta}_{\tau_j}(t)).\end{aligned}$$

Thus,

$$\frac{1}{2}A(t)^2 = \lim_{j\to\infty} \frac{1}{2}\|\zeta_{\tau_j}(t)\|_*^2 \leq \frac{1}{2}\|\xi(t)\|_*^2$$

for almost every $t \geq 0$. We conclude that $A(t) = \|\xi(t)\|_*$ for almost every $t \geq 0$. □

REMARK 4.17. An immediate corollary of the above assertion is that

$$[0,\infty) \ni t \mapsto \|\xi(t)\|_* \text{ is essentially nonincreasing.}$$

4.4. Proof of the main theorem. So far we have $\xi : [0,\infty) \to \mathcal{P}^\perp$, which is our candidate for the weak solution asserted to exist in Theorem 4.3. We still need to establish

$$\mathbf{g}(t) \in \mathcal{J}^*(\xi(t))$$

for almost every $t \geq 0$,

$$[0,\infty) \ni t \mapsto E(\mathbf{g}(t)) \quad \text{is essentially nonincreasing,}$$

and

$$\frac{1}{2}\|\xi(s)\|_*^2 - \frac{1}{2}\|\xi(t)\|_*^2 = E^*V(\xi, s, t) + \int_s^t E(\mathbf{g}(\tau))d\tau$$

for almost every $0 \leq s \leq t$. We'll start with verifying the following assertion.

LEMMA 4.18. *For almost every $t \geq 0$, $\mathbf{g}(t) \in \mathcal{J}^*(\xi(t))$.*

PROOF. For $\mu \in \mathcal{P}^\perp$ and $T > 0$,

$$\begin{aligned}
\int_0^T \frac{1}{2}\|\mu\|_*^2 dt &\geq \int_0^T \left(\frac{1}{2}\|\xi_{\tau_j}(t)\|_*^2 + \langle \mu - \xi_{\tau_j}(t), g_{\tau_j}(t)\rangle\right) dt \\
&= \int_0^T \frac{1}{2}\|\xi_{\tau_j}(t)\|_*^2 dt + \int_0^T \langle \mu - \xi(t), g_{\tau_j}(t)\rangle dt + \int_0^T \langle \xi(t) - \xi_{\tau_j}(t), g_{\tau_j}(t)\rangle dt \\
&\geq \int_0^T \frac{1}{2}\|\xi_{\tau_j}(t)\|_*^2 dt + \int_0^T \langle \mu - \xi(t), g_{\tau_j}(t)\rangle dt - \int_0^T \chi_{\mathcal{C}}^*(\xi(t) - \xi_{\tau_j}(t))dt \\
&= \int_0^T \frac{1}{2}\|\xi(t)\|_*^2 dt + \int_0^T \langle \mu - \xi(t), g_{\tau_j}(t)\rangle dt + o(1)
\end{aligned}$$

as $j \to \infty$. Therefore, it suffices to verify

$$\lim_{j\to\infty} \int_0^T \langle \mu - \xi(t), g_{\tau_j}(t)\rangle dt = \int_0^T \langle \mu - \xi(t), g(t)\rangle dt. \tag{4.21}$$

Note that $\rho(t) := \mu - \xi(t) \in \mathcal{P}^\perp$ satisfies

$$\|\rho(t)\|_* \leq \|\mu\|_* + \|\xi(t)\|_* \leq \|\mu\|_* + \|\xi^0\|_*$$

for almost every $t \geq 0$. Proposition A.2 in the appendix asserts that for each $\epsilon \in (0,1)$, there is $\rho^\epsilon : [0,\infty) \to (\mathcal{V}/\mathcal{P})'$ such that: for all $t \geq 0$, $\rho^\epsilon(t)$ arises as the integration of a smooth function against σ with

$$|\langle \rho^\epsilon(t), g\rangle - \langle \rho(t), g\rangle| \leq 4\sqrt{2}\epsilon\|\rho(t)\|_* \tag{4.22}$$

for each $g \in \mathcal{C}$. Moreover, $t \mapsto \langle \rho^\epsilon(t), h(t)\rangle$ is measurable for any measurable $\boldsymbol{h} : [0,\infty) \to \mathcal{V}/\mathcal{P}$; this is verified in Lemma A.3. Therefore,

$$\left|\int_0^T \langle \rho^\epsilon(t), g_{\tau_j}(t)\rangle dt - \int_0^T \langle \rho(t), g_{\tau_j}(t)\rangle dt\right| \leq 4\sqrt{2}\epsilon T(\|\mu\|_* + \|\xi^0\|_*).$$

We can also solve the PDE

$$-(\Delta h + 2h) = \rho^\epsilon(t)$$

weakly in $\mathbb{S}^2$ for each $t \geq 0$ and $\epsilon \in (0,1)$. For example, we can minimize the functional

$$h \mapsto \int_{\mathbb{S}^2} \left(\frac{1}{2}|\nabla h|^2 - h^2\right) d\sigma - \int_{\mathbb{S}^2} \rho^\epsilon(t) h d\sigma$$

uniquely among $h \in \mathcal{V}$ which satisfies

$$\int_{\mathbb{S}^2} h u_i d\sigma = 0$$

for $i = 1, 2, 3$. The minimizing $h^\epsilon(t) \in \mathcal{V}$ satisfies

$$(\boldsymbol{h}^\epsilon(t), \mathbf{f}) = \int_{\mathbb{S}^2} (\nabla h^\epsilon(t) \cdot \nabla f - 2h^\epsilon(t) f)\, d\sigma = \int_{\mathbb{S}^2} \rho^\epsilon(t) f d\sigma$$

for all $\mathbf{f} \in \mathcal{V}/\mathcal{P}$. Also note that Pettis' theorem (Chapter V section 4 of [**54**]) and Lemma A.3 imply $\boldsymbol{h}^\epsilon : [0, \infty) \to \mathcal{V}/\mathcal{P}$ is measurable.

Observe

$$\begin{aligned}\int_0^T \langle \rho^\epsilon(t), g_{\tau_j}(t)\rangle dt &= \int_0^T \int_{\mathbb{S}^2} \rho^\epsilon(t) g_{\tau_j}(t) d\sigma dt \\ &= \int_0^T \int_{\mathbb{S}^2} \left(\nabla h^\epsilon(t) \cdot \nabla g_{\tau_j}(t) - 2h^\epsilon(t) g_{\tau_j}(t)\right) d\sigma dt.\end{aligned}$$

By the weak convergence of $(\mathbf{g}_{\tau_j})_{j\in\mathbb{N}}$ to $\mathbf{g}$ in $L^2([0,\infty); \mathcal{V}/\mathcal{P})$,

$$\begin{aligned}\lim_{j\to\infty} \int_0^T \langle \rho^\epsilon(t), g_{\tau_j}(t)\rangle dt &= \int_0^T \int_{\mathbb{S}^2} (\nabla h^\epsilon(t) \cdot \nabla g(t) - 2h^\epsilon(t) g(t))\, d\sigma dt \\ &= \int_0^T \int_{\mathbb{S}^2} \rho^\epsilon(t) g(t) d\sigma dt \\ &= \int_0^T \langle \rho^\epsilon(t), g(t)\rangle dt.\end{aligned}$$

Therefore,

$$\begin{aligned}&\limsup_{j\to\infty} \left| \int_0^T \langle \rho(t), g_{\tau_j}(t)\rangle dt - \int_0^T \langle \rho(t), g(t)\rangle dt \right| \\ &\quad \le \limsup_{j\to\infty} \left| \int_0^T \langle \rho(t), g_{\tau_j}(t)\rangle dt - \int_0^T \langle \rho^\epsilon(t), g_{\tau_j}(t)\rangle dt \right| + \\ &\quad\quad + \limsup_{j\to\infty} \left| \int_0^T \langle \rho^\epsilon(t), g_{\tau_j}(t)\rangle dt - \int_0^T \langle \rho(t), g(t)\rangle dt \right| \\ &\quad \le 4\sqrt{2\epsilon} T(\|\mu\|_* + \|\xi^0\|_*) + \left| \int_0^T \langle \rho^\epsilon(t), g(t)\rangle dt - \int_0^T \langle \rho(t), g(t)\rangle dt \right|.\end{aligned}$$

Again recalling that $g(t) \in \mathcal{C}$ for all $t \ge 0$ and using inequality (4.22) leads to

$$\limsup_{j\to\infty} \left| \int_0^T \langle \rho(t), g_{\tau_j}(t)\rangle dt - \int_0^T \langle \rho(t), g(t)\rangle dt \right| \le 8\sqrt{2\epsilon} T(\|\mu\|_* + \|\xi^0\|_*).$$

Since $\epsilon \in (0, 1)$ was arbitrary, we conclude (4.21). □

Next we claim that an energy inequality holds.

LEMMA 4.19. *For almost every* $0 \le s \le t$,

$$\frac{1}{2}\|\xi(s)\|_*^2 - \frac{1}{2}\|\xi(t)\|_*^2 \ge E^* V(\xi, s, t) + \int_s^t E(\mathbf{g}(\tau)) d\tau.$$

PROOF. Assume $0 < s < t$ and choose $0 < \delta < s$. Further select $\tau < \min\{\delta, t - s\}$. Since $s + \tau < t$, we may select $j, k \in \mathbb{N}$ with $j < k$ such that $(j-1)\tau < s \le j\tau$ and $(k-1)\tau < t \le k\tau$. Likewise, since $\tau < \delta$, $s - \delta < (j-1)\tau$; so there is $i \le j-1$ for which $(i-1)\tau < s - \delta \le i\tau$. By (4.9),

$$\begin{aligned}\frac{1}{2}\|\zeta_\tau(s-\delta)\|_*^2 &= \frac{1}{2}\|\xi^i\|_*^2\\ &\ge \frac{1}{2}\|\xi^{j-1}\|_*^2\\ &\ge \frac{1}{2}\|\xi^k\|_*^2 + \sum_{\ell=j}^{k}\tau\left[E^*\left(\frac{\xi^\ell - \xi^{\ell-1}}{\tau}\right) + E(\mathbf{g}^\ell)\right]\\ &= \frac{1}{2}\|\zeta_\tau(t)\|_*^2 + E^*V(\xi_\tau, (j-1)\tau, k\tau) + \int_{(j-1)\tau}^{k\tau} E(\mathbf{g}_\tau(r))dr\\ &\ge \frac{1}{2}\|\zeta_\tau(t)\|_*^2 + E^*V(\xi_\tau, s, t) + \int_s^t E(\mathbf{g}_\tau(r))dr.\end{aligned}$$

We can then pass to the limit along an appropriate sequence of τ tending to 0 to find

$$A(s-\delta) \ge A(t) + E^*V(\xi, s, t) + \int_s^t E(\mathbf{g}(r))dr.$$

Since A is monotone, it is continuous except for possibly on a countable set of times. Therefore, we may send δ to 0 and find

$$A(s) \ge A(t) + E^*V(\xi, s, t) + \int_s^t E(\mathbf{g}(r))dr$$

for all but countably many times $0 \le s \le t$. We conclude upon recalling that $A(t) = \frac{1}{2}\|\xi(t)\|_*^2$ for almost every time $t \ge 0$. □

Now we are in position to establish the monotonicity of $E \circ \mathbf{g}$.

LEMMA 4.20. *The function $E \circ \mathbf{g}$ is essentially nonincreasing.*

PROOF. By the previous lemma,

$$\frac{1}{2}\|\xi(s)\|_*^2 - \frac{1}{2}\|\xi(t)\|_*^2 \ge E^*V(\xi, s, t) + \int_s^t E(\mathbf{g}(\tau))d\tau \tag{4.23}$$

for almost all $0 \le s \le t$. We also have that $\mathbf{g}(s) \in \mathcal{J}^*(\xi(s))$ and $g(s) \in \mathcal{C}$ for almost every $s \ge 0$. For t larger than such an s,

$$\begin{aligned}\frac{1}{2}\|\xi(t)\|_*^2 &\ge \frac{1}{2}\|\xi(s)\|_*^2 + \langle \xi(t) - \xi(s), g(s)\rangle\\ &\ge \frac{1}{2}\|\xi(s)\|_*^2 + (t-s)\left\langle \frac{\xi(t)-\xi(s)}{t-s}, g(s)\right\rangle\\ &\ge \frac{1}{2}\|\xi(s)\|_*^2 - (t-s)E^*\left(\frac{\xi(t)-\xi(s)}{t-s}\right) - (t-s)E(\mathbf{g}(s))\\ &\ge \frac{1}{2}\|\xi(s)\|_*^2 - E^*V(\xi, s, t) - (t-s)E(\mathbf{g}(s)). \qquad (4.24)\end{aligned}$$

That is,

$$\frac{1}{2}\|\xi(s)\|_*^2 - \frac{1}{2}\|\xi(t)\|_*^2 \le E^*V(\xi, s, t) + (t-s)E(\mathbf{g}(s)). \tag{4.25}$$

Comparing (4.23) and (4.25), we find

$$\int_s^t E(\mathbf{g}(\tau))d\tau \le (t-s)E(\mathbf{g}(s))$$

for almost all $0 \le s \le t$. By Lemma B.3 in the appendix, $E \circ \mathbf{g}$ is necessarily essentially nonincreasing. □

The final detail needed in our proof of Theorem 4.3 is that equality (essentially) holds in the energy identity.

LEMMA 4.21. *For almost every $s \le t$,*

$$\frac{1}{2}\|\xi(s)\|_*^2 - \frac{1}{2}\|\xi(t)\|_*^2 = E^*V(\xi, s, t) + \int_s^t E(\mathbf{g}(\tau))d\tau.$$

PROOF. By Lemma B.1, we may choose a nonincreasing function $I : [0, \infty) \to [0, \infty)$ such that

$$I(t) = E(\mathbf{g}(t))$$

for almost every $t \ge 0$. Let $0 \le s < t$ be such that $I(s) = E(\mathbf{g}(s))$, $I(t) = E(\mathbf{g}(t))$, and $\mathbf{g}(s) \in \mathcal{J}^*(\xi(s))$. Since I is nonincreasing and $0 \le I(\tau) \le E(\mathbf{g}^0)$ for almost every $\tau \ge 0$, I is Riemann integrable on $[s, t]$.

Fix $\epsilon > 0$. We may select $\delta > 0$ such that for any partition $s = t_0 < t_1 < \cdots < t_N = t$ with $\max_{1 \le i \le N}(t_i - t_{i-1}) \le \delta$ implies

$$\sum_{i=1}^N (t_i - t_{i-1})(I(t_{i-1}) - I(t_i)) \le \epsilon.$$

In view of Lemma 4.18, we may choose such a partition for which $\mathbf{g}(t_i) \in \mathcal{J}^*(\xi(t_i))$ and $I(t_i) = E(\mathbf{g}(t_i))$ for $i = 1, \dots, N-1$. Employing inequality (4.24), we find

$$\begin{aligned}
\frac{1}{2}\|\xi(s)\|_*^2 - \frac{1}{2}\|\xi(t)\|_*^2 &= \sum_{i=1}^N \left[\frac{1}{2}\|\xi(t_{i-1})\|_*^2 - \frac{1}{2}\|\xi(t_i)\|_*^2\right] \\
&\le \sum_{i=1}^N [E^*V(\xi, t_{i-1}, t_i) + (t_i - t_{i-1})I(t_{i-1})] \\
&= E^*V(\xi, s, t) + \sum_{i=1}^N (t_i - t_{i-1})I(t_{i-1}) \\
&\le E^*V(\xi, s, t) + \sum_{i=1}^N (t_i - t_{i-1})I(t_i) + \epsilon \\
&\le E^*V(\xi, s, t) + \int_s^t I(\tau)d\tau + \epsilon \\
&= E^*V(\xi, s, t) + \int_s^t E(\mathbf{g}(\tau))d\tau + \epsilon.
\end{aligned}$$

Since ϵ was arbitrary,

$$\frac{1}{2}\|\xi(s)\|_*^2 - \frac{1}{2}\|\xi(t)\|_*^2 \le E^*V(\xi, s, t) + \int_s^t E(\mathbf{g}(\tau))d\tau.$$

We conclude by noting that this inequality and the opposite inequality holds for almost every $0 \le s \le t$. □

Appendix A. Smoothing elements of $\mathcal{P}^\perp$

We will discuss a way to smooth measures on $\mathbb{S}^2$ which is tailored for the needs of this paper. With this goal in mind, we will choose a family $(\psi^\epsilon)_{\epsilon\in(0,1)}$ satisfying

$$\begin{cases} \psi^\epsilon \in C^\infty(\mathbb{R}) \\ \psi^\epsilon \geq 0 \\ \operatorname{supp}(\psi^\epsilon) = [1-\epsilon, 1] \end{cases}$$

and

$$\int_{\mathbb{S}^2} \psi^\epsilon(u_1) d\sigma(u) = 1$$

for each $\epsilon \in (0,1)$. Since σ is invariant under orthogonal transformations,

$$\int_{\mathbb{S}^2} \psi^\epsilon(u\cdot v) d\sigma(u) = 1$$

for each $v \in \mathbb{S}^2$. Moreover, the support of $u \mapsto \psi^\epsilon(u \cdot v)$ is the spherical cap $\{u \in \mathbb{S}^2 : u\cdot v \geq 1-\epsilon\}$.

For a given $\mu \in M(\mathbb{S}^2)$, we can define

$$\tilde{\mu}^\epsilon(u) := \int_{\mathbb{S}^2} \psi^\epsilon(u\cdot v) d\mu(v) \tag{A.1}$$

for $u \in \mathbb{S}^2$. It is routine to check that $\tilde{\mu}^\epsilon \in C^\infty(\mathbb{S}^2)$. We will identify this function with the measure it induces when integrated against σ:

$$\int_{\mathbb{S}^2} g d\tilde{\mu}^\epsilon := \int_{\mathbb{S}^2} g(u)\tilde{\mu}^\epsilon(u) d\sigma(u) = \int_{\mathbb{S}^2}\int_{\mathbb{S}^2} g(u)\psi^\epsilon(u\cdot v) d\mu(v) d\sigma(u)$$

for $g \in C(\mathbb{S}^2)$.

LEMMA A.1. *Suppose* $\mu \in M(\mathbb{S}^2)$ *and* $g \in C(\mathbb{S}^2)$ *satisfies*

$$|g(u) - g(v) \leq |u - v|$$

for $u, v \in \mathbb{S}^2$. *Then*

$$\left|\int_{\mathbb{S}^2} g d\tilde{\mu}^\epsilon - \int_{\mathbb{S}^2} g d\mu\right| \leq \sqrt{2\epsilon}\, \|\mu\|_*. \tag{A.2}$$

PROOF. Observe

$$\begin{aligned}
\int_{\mathbb{S}^2} g d\tilde{\mu}^\epsilon - \int_{\mathbb{S}^2} g d\mu &= \int_{\mathbb{S}^2}\int_{\mathbb{S}^2} g(u)\psi^\epsilon(u\cdot v) d\mu(v) d\sigma(u) - \int_{\mathbb{S}^2} g(v) d\mu(v) \\
&= \int_{\mathbb{S}^2} \left[\int_{\mathbb{S}^2} g(u)\psi^\epsilon(u\cdot v) d\sigma(u) - g(v)\right] d\mu(v) \\
&= \int_{\mathbb{S}^2} \left[\int_{\mathbb{S}^2} (g(u) - g(v))\psi^\epsilon(u\cdot v) d\sigma(u)\right] d\mu(v) \\
&= \int_{\mathbb{S}^2} \left[\int_{\{u\cdot v\geq 1-\epsilon\}} (g(u) - g(v))\psi^\epsilon(u\cdot v) d\sigma(u)\right] d\mu(v) \\
&\leq \int_{\mathbb{S}^2} \left[\int_{\{u\cdot v\geq 1-\epsilon\}} |g(u) - g(v)|\psi^\epsilon(u\cdot v) d\sigma(u)\right] d|\mu|(v) \\
&\leq \int_{\mathbb{S}^2} \left[\int_{\{u\cdot v\geq 1-\epsilon\}} |u - v|\psi^\epsilon(u\cdot v) d\sigma(u)\right] d|\mu|(v).
\end{aligned}$$

Note that on the spherical cap $\{u \in \mathbb{S}^2 : u \cdot v \geq 1 - \epsilon\}$,

$$|u - v|^2 = 2(1 - u \cdot v) \leq 2\epsilon.$$

As a result,

$$\int_{\mathbb{S}^2} g d\tilde{\mu}^\epsilon - \int_{\mathbb{S}^2} g d\mu \leq \sqrt{2\epsilon} \int_{\mathbb{S}^2} \left[\int_{\mathbb{S}^2} \psi^\epsilon(u \cdot v) d\sigma(u) \right] d|\mu|(v) = \sqrt{2\epsilon} \|\mu\|_*.$$

We can use a similar argument to obtain the same upper bound for $\int_{\mathbb{S}^2} g d\mu - \int_{\mathbb{S}^2} g d\tilde{\mu}^\epsilon$. □

We will now introduce the approximation

$$\mu^\epsilon(u) := \tilde{\mu}^\epsilon(u) - a^\epsilon \cdot u, \tag{A.3}$$

where $a^\epsilon \in \mathbb{R}^3$ is chosen so that μ^ϵ is orthogonal to $\mathcal{P}$ in $L^2(\mathbb{S}^2)$. That is,

$$a_i^\epsilon = \frac{\int_{\mathbb{S}^2} u_i \tilde{\mu}^\epsilon(u) d\sigma(u)}{\int_{\mathbb{S}^2} u_i^2 d\sigma(u)}, \quad i = 1, 2, 3.$$

PROPOSITION A.2. *Suppose $g \in \mathcal{C}$, $\mu \in \mathcal{P}^\perp$, and define μ^ϵ by* (A.3). *Then $\mu^\epsilon \in C^\infty(\mathbb{S}^2)$, as a measure $\mu^\epsilon \in \mathcal{P}^\perp$, and*

$$\left| \int_{\mathbb{S}^2} g d\mu^\epsilon - \int_{\mathbb{S}^2} g d\mu \right| \leq 4\sqrt{2\epsilon}\, \|\mu\|_*.$$

PROOF. It is clear that $\mu^\epsilon \in C^\infty(\mathbb{S}^2)$. Since μ^ϵ is orthogonal to $\mathcal{P}$ in $L^2(\mathbb{S}^2)$, the measure it induces belongs to $\mathcal{P}^\perp$ as

$$\int_{\mathbb{S}^2} u_i d\mu^\epsilon(u) = \int_{\mathbb{S}^2} u_i \mu^\epsilon(u) d\sigma(u) = 0$$

$(i = 1, 2, 3)$.

Now let H be the support function of a constant width body K associated with g. Let $a \in K$ be the center of the circumball of K. By Corollary 2.8,

$$\begin{cases} |H(u) - 1/2 - a \cdot u| \leq 1 \\ |H(u) - H(v) - a \cdot (u - v)| \leq |u - v| \end{cases}$$

for all $u, v \in \mathbb{S}^2$. Therefore, if we set $\tilde{g}(u) = g(u) - a \cdot u$,

$$\begin{cases} |\tilde{g}(u)| \leq 1 \\ |\tilde{g}(u) - \tilde{g}(v)| \leq |u - v| \end{cases}$$

for all $u, v \in \mathbb{S}^2$.

By inequality (A.2),

$$\begin{aligned} \left| \int_{\mathbb{S}^2} g d\mu^\epsilon - \int_{\mathbb{S}^2} g d\mu \right| &= \left| \int_{\mathbb{S}^2} \tilde{g} d\mu^\epsilon - \int_{\mathbb{S}^2} \tilde{g} d\mu \right| \\ &= \left| \int_{\mathbb{S}^2} \tilde{g} d\tilde{\mu}^\epsilon - \int_{\mathbb{S}^2} \tilde{g} d\mu - \int_{\mathbb{S}^2} \tilde{g}(u) a^\epsilon \cdot u d\sigma(u) \right| \\ &\leq \sqrt{2\epsilon} \|\mu\|_* + \left| \int_{\mathbb{S}^2} \tilde{g}(u) a^\epsilon \cdot u d\sigma(u) \right| \\ &\leq \sqrt{2\epsilon} \|\mu\|_* + 4\pi |a^\epsilon|. \end{aligned}$$

Recall

$$\int_{\mathbb{S}^2} u_i^2 d\sigma(u) = \frac{4\pi}{3}$$

for $i = 1, 2, 3$ and $\mu \in \mathcal{P}^\perp$. It follows that

$$\begin{aligned}
\frac{4\pi}{3} a^\epsilon &= \int_{\mathbb{S}^2} u\tilde{\mu}^\epsilon(u) d\sigma(u) \\
&= \int_{\mathbb{S}^2}\int_{\mathbb{S}^2} u\psi^\epsilon(u\cdot v) d\mu(v) d\sigma(u) \\
&= \int_{\mathbb{S}^2}\int_{\mathbb{S}^2} (u-v)\psi^\epsilon(u\cdot v) d\sigma(u) d\mu(v) + \int_{\mathbb{S}^2}\int_{\mathbb{S}^2} v\psi^\epsilon(u\cdot v) d\mu(v) d\sigma(u) \\
&= \int_{\mathbb{S}^2}\int_{\mathbb{S}^2} (u-v)\psi^\epsilon(u\cdot v) d\sigma(u) d\mu(v) + \int_{\mathbb{S}^2} v \left(\int_{\mathbb{S}^2} \psi^\epsilon(u\cdot v) d\sigma(u)\right) d\mu(v) \\
&= \int_{\mathbb{S}^2}\int_{\mathbb{S}^2} (u-v)\psi^\epsilon(u\cdot v) d\sigma(u) d\mu(v) + \int_{\mathbb{S}^2} v d\mu(v) \\
&= \int_{\mathbb{S}^2}\int_{\mathbb{S}^2} (u-v)\psi^\epsilon(u\cdot v) d\sigma(u) d\mu(v).
\end{aligned}$$

Therefore,

$$\begin{aligned}
4\pi|a^\epsilon| &\le 3\int_{\mathbb{S}^2}\int_{\mathbb{S}^2} |u-v|\psi^\epsilon(u\cdot v) d\sigma(u) d|\mu|(v) \\
&= 3\int_{\mathbb{S}^2}\int_{u\cdot v\ge 1-\epsilon} |u-v|\psi^\epsilon(u\cdot v) d\sigma(u) d|\mu|(v) \\
&\le 3\sqrt{2\epsilon}\int_{\mathbb{S}^2}\int_{\mathbb{S}^2} \psi^\epsilon(u\cdot v) d\sigma(u) d|\mu|(v) \\
&= 3\sqrt{2\epsilon}\|\mu\|_*.
\end{aligned}$$

Consequently,

$$\left|\int_{\mathbb{S}^2} g d\mu^\epsilon - \int_{\mathbb{S}^2} g d\mu\right| \le \sqrt{2\epsilon}\|\mu\|_* + 3\sqrt{2\epsilon}\|\mu\|_*.$$

□

We will need one more technical assertion regarding how this smoothing technique is applied to measurable mappings $\rho : [0,\infty) \to \mathcal{P}^\perp$. We will employ the fact that if $\mu \in \mathcal{P}^\perp$, then

$$\left|\int_{\mathbb{S}^2} \mu^\epsilon h d\sigma\right| \le \frac{1}{2}\left(\int_{\mathbb{S}^2} (\mu^\epsilon)^2 d\sigma\right)^{1/2} (\boldsymbol{h},\boldsymbol{h})^{1/2} \tag{A.4}$$

for all $\boldsymbol{h} \in \mathcal{V}/\mathcal{P}$; we established this inequality in our proof of Lemma 3.5. In particular,

$$\mathcal{V}/\mathcal{P} \ni \boldsymbol{h} \mapsto \int_{\mathbb{S}^2} h\mu^\epsilon d\sigma$$

is continuous, so we may consider μ^ϵ an element of $(\mathcal{V}/\mathcal{P})'$.

LEMMA A.3. *Suppose $\rho : [0,\infty) \to \mathcal{P}^\perp$ is measurable and define $\rho^\epsilon : [0,\infty) \to (\mathcal{V}/\mathcal{P})'$ by*

$$\rho^\epsilon(t) := (\rho(t))^\epsilon$$

for each $t \geq 0$. *Then for all* $\boldsymbol{h} \in \mathcal{V}/\mathcal{P}$,

$$[0,\infty) \ni t \mapsto \int_{\mathbb{S}^2} h\rho^\epsilon(t)d\sigma$$

is measurable. Moreover, for all measurable $\mathbf{g} : [0,\infty) \to \mathcal{V}/\mathcal{P}$,

$$[0,\infty) \ni t \mapsto \int_{\mathbb{S}^2} g(t)\rho^\epsilon(t)d\sigma$$

is measurable.

PROOF. Let $\boldsymbol{h} \in \mathcal{V}/\mathcal{P}$. For a given $t \geq 0$, set $\mu = \rho(t)$ and define $\tilde{\rho}^\epsilon(t) := \tilde{\mu}^\epsilon$ as in (A.1). Note that

$$\int_{\mathbb{S}^2} h\tilde{\rho}^\epsilon(t)d\sigma = \langle \rho(t), h^\epsilon \rangle,$$

where

$$h^\epsilon(u) := \int_{\mathbb{S}^2} \psi^\epsilon(u \cdot v)g(v)d\sigma(v)$$

for $u \in \mathbb{S}^2$. We also have

$$\begin{aligned}\int_{\mathbb{S}^2} \rho^\epsilon(t)hd\sigma &= \int_{\mathbb{S}^2} h\tilde{\rho}^\epsilon(t)d\sigma - \int_{\mathbb{S}^2} a^\epsilon(t)\cdot uh(u)d\sigma(u)\\ &= \langle \rho(t), h^\epsilon\rangle - \sum_{i=1}^3 a_i^\epsilon(t)\left(\int_{\mathbb{S}^2} u_i h(u)d\sigma(u)\right),\end{aligned}$$

where

$$a_i^\epsilon(t) = \frac{\int_{\mathbb{S}^2} u_i\tilde{\rho}^\epsilon(t)d\sigma}{\int_{\mathbb{S}^2}(u_i)^2 d\sigma} = \langle \rho^\epsilon(t), \phi_i\rangle = \langle \rho(t), \phi_i^\epsilon\rangle$$

for $i = 1, 2, 3$. Here $\phi_i(v) = v_i / \int_{\mathbb{S}^2}(u_i)^2 d\sigma$. Thus,

$$\int_{\mathbb{S}^2} \rho^\epsilon(t)hd\sigma = \langle \rho(t), h^\epsilon\rangle - \sum_{i=1}^3 \left(\int_{\mathbb{S}^2} u_i h(u)d\sigma(u)\right)\langle \rho(t), \phi_i^\epsilon\rangle$$

is a measurable function since ρ is measurable.

Observe that for any fixed t,

$$\mathcal{V}/\mathcal{P} \ni \boldsymbol{h} \mapsto \int_{\mathbb{S}^2} h\rho^\epsilon(t)d\sigma$$

is continuous. This continuity follows directly from inequality (A.4). Therefore,

$$[0,\infty) \times \mathcal{V}/\mathcal{P} \ni (t, \boldsymbol{h}) \mapsto \int_{\mathbb{S}^2} h\rho^\epsilon(t)d\sigma$$

is a Carathéodory function. As a result, if $\mathbf{g} : [0,\infty) \to \mathcal{V}/\mathcal{P}$ is measurable, so is

$$[0,\infty) \ni t \mapsto \int_{\mathbb{S}^2} g(t)\rho^\epsilon(t)d\sigma$$

(Lemma 8.2.3 of [**6**]). □

Appendix B. Essentially monotone functions

We'll say that a function $f : [0, \infty) \to \mathbb{R}$ is *essentially nondecreasing* if there is a set $N \subset [0, \infty)$ of measure 0 for which $f(t) \geq f(s)$ whenever $t \geq s$ and $t, s \notin N$. If under the same conditions, $f(t) \leq f(s)$, we'll say f is *essentially nonincreasing.* If f is either essentially nondecreasing or nonincreasing, we'll say f is *essentially monotone.*

It turns out that the only way essentially monotone functions arise is by altering monotone functions on sets of measure 0.

LEMMA B.1. *Suppose $f : [0, \infty) \to \mathbb{R}$ is essentially nondecreasing. There is a nondecreasing function $g : [0, \infty) \to [-\infty, \infty)$ such that $f(t) = g(t)$ for almost every $t \geq 0$.*

PROOF. Choose a null set $N \subset [0, \infty)$ such that $f|_{N^c}$ is nondecreasing. For $t \geq 0$, let us also define

$$g(t) := \sup \{c \in \mathbb{R} : c \leq f(\tau),\ \tau \geq t, \tau \notin N\}.$$

Suppose $t \notin N$. Then $c \leq f(t)$ in the definition above as we may choose $\tau = t$; therefore, $g(t) \leq f(t)$. And as $f(t) \leq f(\tau)$ for $\tau \geq t$ and $\tau \notin N$, we also have $f(t) \leq g(t)$. Thus, f and g agree on N^c.

Let us check that g is nondecreasing. Suppose $t_1 \leq t_2$. Observe that if $c \leq f(\tau)$ for all $\tau \geq t_1$ with $\tau \notin N$, then $c \leq f(\tau)$ for all $\tau \geq t_2$ with $\tau \notin N$. In this case, $c \leq g(t_2)$. Therefore, $g(t_1) \leq g(t_2)$. Also note that for any $t \geq 0$, there is $\tau > t$ with $\tau \notin N$. Therefore, $g(t) \leq g(\tau) = f(\tau) < \infty$. □

We now generalize an elementary fact: a Riemann integrable function from $[0, \infty)$ into $[0, \infty)$ which is nonincreasing tends to 0 faster than $1/t$ as $t \to \infty$.

PROPOSITION B.2. *Suppose that $f : [0, \infty) \to [0, \infty)$ is essentially nonincreasing and*

$$\int_0^\infty f(t)dt < \infty.$$

Then there is a null set $N \subset [0, \infty)$ for which

$$\lim_{\substack{t \to \infty \\ t \notin N}} tf(t) = 0. \tag{B.1}$$

PROOF. Let $N \subset [0, \infty)$ be a null set for which $f|_{N^c}$ is nonincreasing. Next, choose an unbounded, increasing sequence of times $(t_k)_{k \in \mathbb{N}} \subset N^c$ and fix $\epsilon > 0$. Since f is integrable, there is $K(\epsilon) \in \mathbb{N}$ such that

$$\frac{t_k}{2} f(t_k) \leq \int_{t_k/2}^{t_k} f(t)dt \leq \frac{1}{2}\epsilon$$

for $k \geq K(\epsilon)$. That is, $t_k f(t_k) \leq \epsilon$ for $k \geq K(\epsilon)$. Therefore, $\lim_{k \to \infty} t_k f(t_k) = 0$. We conclude (B.1). □

Note that if $f : [0, \infty) \to \mathbb{R}$ is essentially nonincreasing, then

$$\int_s^t f(\tau)d\tau \leq (t - s)f(s) \tag{B.2}$$

for almost every $0 \leq s \leq t < \infty$. The converse is also true.

LEMMA B.3. *Suppose $f : [0,\infty) \to \mathbb{R}$ is locally integrable and* (B.2) *holds for almost every $0 \le s \le t < \infty$. Then f is essentially nonincreasing.*

PROOF. 1. First suppose that f is smooth in $(0,\infty)$. This implies f satisfies (B.2) for all $0 < s \le t < \infty$. We can integrate by parts to find

$$\begin{aligned}\int_s^t f(\tau)d\tau &= \int_s^t f(\tau)\frac{d}{d\tau}(\tau - t)d\tau \\ &= (\tau - t)f(\tau)\Big|_s^t - \int_s^t f'(\tau)(\tau - t)d\tau \\ &= f(s)(t-s) + \int_s^t f'(\tau)(t-\tau)d\tau \\ &\le f(s)(t-s).\end{aligned}$$

Consequently,

$$\int_s^t f'(\tau)(t-\tau)d\tau \le 0 \tag{B.3}$$

for all $0 < s \le t < \infty$.

We claim that $f'(\tau) \le 0$ for all $0 < \tau < \infty$. Otherwise, $f'(s_0) > 0$ for some $s_0 \in (0,\infty)$. By continuity, there is $\delta > 0$ for which

$$f'(\tau) \ge \frac{1}{2}f'(s_0)$$

for $s_0 \le \tau < s_0 + \delta$. In particular,

$$\int_{s_0}^{s_0+\delta} f'(\tau)((s_0+\delta) - \tau)d\tau \ge \frac{1}{2}f'(s_0)\cdot \int_{s_0}^{s_0+\delta}((s_0+\delta)-\tau)d\tau = \frac{1}{2}f'(s_0)\cdot\frac{\delta^2}{2} > 0.$$

This would contradict (B.3) with $s = s_0$ and $t = s_0 + \delta$. As a result, if f is smooth and satisfies (B.2), then f is necessarily nonincreasing.

2. Now let us consider general f and set $f^\epsilon := \eta^\epsilon * f$ as the standard mollification of f which is defined on (ϵ,∞) (Chapter 4 of [**21**]). Let $\epsilon < s < t < \infty$. Observe that

$$\begin{aligned}\int_s^t f^\epsilon(\tau)d\tau &= \int_s^t \left(\int_{-\epsilon}^{\epsilon} \eta^\epsilon(r)f(\tau - r)dr\right)d\tau \\ &= \int_{-\epsilon}^{\epsilon} \eta^\epsilon(r)\left(\int_s^t f(\tau - r)d\tau\right)dr \\ &= \int_{-\epsilon}^{\epsilon} \eta^\epsilon(r)\left(\int_{s-r}^{t-r} f(\rho)d\rho\right)dr.\end{aligned}$$

As (B.2) holds outside of a null set,

$$\int_{s-r}^{t-r} f(\rho)d\rho \le ((t-r)-(s-t))f(s-r) = (t-s)f(s-r)$$

for almost every $|r| < \epsilon$. Therefore,

$$\int_s^t f^\epsilon(\tau)d\tau \le (t-s)\int_{-\epsilon}^{\epsilon}\eta^\epsilon(r)f(s-r)dr = (t-s)f^\epsilon(s).$$

By part 1 of this proof, f^ϵ is nonincreasing on (ϵ,∞). Since $f^\epsilon(s) \to f(s)$ for almost every $s \in (0,\infty)$, we have that f is essentially nonincreasing. $\square$

Appendix C. Essentially lower-semicontinuous functions

We will say that a function $f : [0, \infty) \to \mathbb{R}$ is *essentially lower-semicontinuous* if f is lower-semicontinuous at t for almost every $t \geq 0$. A typical lower-semicontinous function $g : [0, \infty) \to \mathbb{R}$ is Borel measurable as its sublevel set $g^{-1}((-\infty, y])$ is closed in $[0, \infty)$ for each $y \in \mathbb{R}$. It turns out that an essentially lower-semicontinuous function is at least Lebesgue measurable.

PROPOSITION C.1. *Suppose $f : [0, \infty) \to \mathbb{R}$ is essentially lower-semicontinuous. Then f is Lebesgue measurable.*

PROOF. Let $N \subset [0, \infty)$ be a null set for which f may not be lower-semicontinuous at times $t \in N$. For a given $y \in \mathbb{R}$, we claim that

$$f^{-1}((y, \infty)) \cap N^c = O \cap N^c \tag{C.1}$$

for some open $O \subset \mathbb{R}$. If we can establish this claim, then

$$\begin{aligned} f^{-1}((y, \infty)) &= f^{-1}((y, \infty)) \cap (N \cup N^c) \\ &= \left(f^{-1}((y, \infty)) \cap N\right) \cup \left(f^{-1}((y, \infty)) \cap N^c\right) \\ &= \left(f^{-1}((y, \infty)) \cap N\right) \cup (O \cap N^c) \\ &= \left(f^{-1}((y, \infty)) \cap N\right) \cup (O^c \cup N)^c. \end{aligned}$$

Since $f^{-1}((y, \infty)) \cap N$ is a null set and $O^c \cup N$ is Lebesgue measurable, $f^{-1}((y, \infty))$ would be Lebesgue measurable.

Let us now verify (C.1). Let $t \in f^{-1}((y, \infty)) \cap N^c$. That is, $f(t) > y$ and f is lower-semicontinuous at t. We note $f(t) - \epsilon > y$ for some $\epsilon > 0$. We may also select $\delta_t > 0$ such that if $|t - s| < \delta_t$ and $s \geq 0$, then

$$f(t) \leq f(s) + \epsilon.$$

It follows that $f(s) \geq f(t) - \epsilon > y$ and thus

$$(t - \delta_t, t + \delta_t) \cap N^c \subset f^{-1}((y, \infty)) \cap N^c. \tag{C.2}$$

Let us choose

$$O := \bigcup \{(t - \delta_t, t + \delta_t) : t \in f^{-1}((y, \infty)) \cap N^c\}$$

and note (C.2) implies

$$O \cap N^c \subset f^{-1}((y, \infty)) \cap N^c.$$

Alternatively, if $t \in f^{-1}((y, \infty)) \cap N^c$, then $t \in (t - \delta_t, t + \delta_t) \cap N^c \subset O \cap N^c$. We conclude $f^{-1}((y, \infty)) \cap N^c \subset O \cap N^c$. This verifies (C.1). □

Acknowledgments

Figures 2 and 3 were made with the Meissner Tetrahedra – Wolfram Demonstration Project. We would also like to thank Kristine Kreidler and Cheikh N'diaye for their interest in this work and for their encouragement.

References

[1] Luigi Ambrosio, Nicola Gigli, and Giuseppe Savaré, *Gradient flows in metric spaces and in the space of probability measures*, 2nd ed., Lectures in Mathematics ETH Zürich, Birkhäuser Verlag, Basel, 2008. MR2401600

[2] Henri Anciaux and Nikos Georgiou, *The Blaschke-Lebesgue problem for constant width bodies of revolution*, 2009.

[3] Henri Anciaux and Brendan Guilfoyle, *On the three-dimensional Blaschke-Lebesgue problem*, Proc. Amer. Math. Soc. **139** (2011), no. 5, 1831–1839, DOI 10.1090/S0002-9939-2010-10588-9. MR2763770

[4] Pedro R. S. Antunes and Beniamin Bogosel, *Parametric Shape Optimization using the Support Function*, 2019.

[5] Tsutomu Arai, *On the existence of the solution for* $\partial\varphi(u'(t)) + \partial\psi(u(t)) \ni f(t)$, J. Fac. Sci. Univ. Tokyo Sect. IA Math. **26** (1979), no. 1, 75–96. MR539774

[6] Jean-Pierre Aubin and Hélène Frankowska, *Set-valued analysis*, Systems & Control: Foundations & Applications, vol. 2, Birkhäuser Boston, Inc., Boston, MA, 1990. MR1048347

[7] Viorel Barbu, *Existence theorems for a class of two point boundary problems*, J. Differential Equations **17** (1975), 236–257, DOI 10.1016/0022-0396(75)90043-1. MR380532

[8] T. Bayen, T. Lachand-Robert, and É. Oudet, *Analytic parametrization of three-dimensional bodies of constant width*, Arch. Ration. Mech. Anal. **186** (2007), no. 2, 225–249, DOI 10.1007/s00205-007-0060-x. MR2342202

[9] Dominique Blanchard, Alain Damlamian, and Hamid Ghidouche, *A nonlinear system for phase change with dissipation*, Differential Integral Equations **2** (1989), no. 3, 344–362. MR983686

[10] Wilhelm Blaschke, *Einige Bemerkungen über Kurven und Flächen konstanter Breite*, Ber. Verh. sächs. Akad. Leipzig **67** (1915), 290–297.

[11] Wilhelm Blaschke, *Konvexe Bereiche gegebener konstanter Breite und kleinsten Inhalts* (German), Math. Ann. **76** (1915), no. 4, 504–513, DOI 10.1007/BF01458221. MR1511839

[12] Giovanna Bonfanti, Michel Frémond, and Fabio Luterotti, *Global solution to a nonlinear system for irreversible phase changes*, Adv. Math. Sci. Appl. **10** (2000), no. 1, 1–24. MR1769184

[13] T. Bonnesen and W. Fenchel, *Theory of convex bodies*, BCS Associates, Moscow, ID, 1987. Translated from the German and edited by L. Boron, C. Christenson and B. Smith. MR920366

[14] Stefano Campi, Andrea Colesanti, and Paolo Gronchi, *Minimum problems for volumes of convex bodies*, Partial differential equations and applications, Lecture Notes in Pure and Appl. Math., vol. 177, Dekker, New York, 1996, pp. 43–55. MR1371579

[15] G. D. Chakerian, *Sets of constant width*, Pacific J. Math. **19** (1966), 13–21. MR205152

[16] P. Colli and A. Visintin, *On a class of doubly nonlinear evolution equations*, Comm. Partial Differential Equations **15** (1990), no. 5, 737–756, DOI 10.1080/03605309908820706. MR1070845

[17] Pierluigi Colli, *On some doubly nonlinear evolution equations in Banach spaces*, Japan J. Indust. Appl. Math. **9** (1992), no. 2, 181–203, DOI 10.1007/BF03167565. MR1170721

[18] Giovanni Colombo and Paolo Gidoni, *On the optimal control of rate-independent soft crawlers* (English, with English and French summaries), J. Math. Pures Appl. (9) **146** (2021), 127–157, DOI 10.1016/j.matpur.2020.11.005. MR4197283

[19] Hallard T. Croft, Kenneth J. Falconer, and Richard K. Guy, *Unsolved problems in geometry*, Problem Books in Mathematics, Springer-Verlag, New York, 1994. Corrected reprint of the 1991 original [MR1107516 (92c:52001)]; Unsolved Problems in Intuitive Mathematics, II. MR1316393

[20] H. G. Eggleston, *A proof of Blaschke's theorem on the Reuleaux triangle*, Quart. J. Math. Oxford Ser. (2) **3** (1952), 296–297, DOI 10.1093/qmath/3.1.296. MR51543

[21] Lawrence C. Evans and Ronald F. Gariepy, *Measure theory and fine properties of functions*, Revised edition, Textbooks in Mathematics, CRC Press, Boca Raton, FL, 2015. MR3409135

[22] Gilles Francfort and Alexander Mielke, *Existence results for a class of rate-independent material models with nonconvex elastic energies*, J. Reine Angew. Math. **595** (2006), 55–91, DOI 10.1515/CRELLE.2006.044. MR2244798

[23] Matsusaburô Fujiwara, *Analytic proof of Blaschke's theorem on the curve of constant breadth with minimum area*, Proc. Imp. Acad. Tokyo **3** (1927), no. 6, 307–309. MR1568234

[24] Mostafa Ghandehari, *An optimal control formulation of the Blaschke-Lebesgue theorem*, J. Math. Anal. Appl. **200** (1996), no. 2, 322–331, DOI 10.1006/jmaa.1996.0208. MR1391153

[25] Evans M. Harrell II, *A direct proof of a theorem of Blaschke and Lebesgue*, J. Geom. Anal. **12** (2002), no. 1, 81–88, DOI 10.1007/BF02930861. MR1881292

[26] Emmanuel Hebey, *Sobolev spaces on Riemannian manifolds*, Lecture Notes in Mathematics, vol. 1635, Springer-Verlag, Berlin, 1996, DOI 10.1007/BFb0092907. MR1481970

[27] Ralph Howard, *Convex bodies of constant width and constant brightness*, Adv. Math. **204** (2006), no. 1, 241–261, DOI 10.1016/j.aim.2005.05.015. MR2233133

[28] I. M. Jaglom and V. G. Boltjanskiĭ, *Convex figures*, Holt, Rinehart and Winston, New York, 1960. Translated by Paul J. Kelly and Lewis F. Walton. MR0123962

[29] Heinrich Jung, *Ueber die kleinste Kugel, die eine räumliche Figur einschliesst* (German), J. Reine Angew. Math. **123** (1901), 241–257, DOI 10.1515/crll.1901.123.241. MR1580570

[30] Bernd Kawohl and Christof Weber, *Meissner's mysterious bodies*, Math. Intelligencer **33** (2011), no. 3, 94–101, DOI 10.1007/s00283-011-9239-y. MR2844102

[31] Pavel Krejčí and Giselle A. Monteiro, *What is the best viscous approximation to a rate-independent process?*, J. Convex Anal. **27** (2020), no. 3, 1015–1032. MR4158534

[32] C. J. Larsen, M. Ortiz, and C. L. Richardson, *Fracture paths from front kinetics: relaxation and rate independence*, Arch. Ration. Mech. Anal. **193** (2009), no. 3, 539–583, DOI 10.1007/s00205-009-0216-y. MR2525111

[33] Henri Lebesgue, *Sur le problème des isopérimètres et sur les domaines de largeur constante*, Bulletin de la Société Mathématique de France C. R. **7** (1914), 72–76.

[34] Mark Kac, *Sur quelques questions de minimum, relatives aux courbes orbiformes, et sur leurs rapports avec le calcul des variations*, J. Math. pures appl. **4** (1921), no. 8, 67–96.

[35] Andreas Mainik and Alexander Mielke, *Existence results for energetic models for rate-independent systems*, Calc. Var. Partial Differential Equations **22** (2005), no. 1, 73–99, DOI 10.1007/s00526-004-0267-8. MR2105969

[36] Federica Malagoli, *An optimal control theory approach to the Blaschke-Lebesgue theorem*, J. Convex Anal. **16** (2009), no. 2, 391–407. MR2559951

[37] Horst Martini, Luis Montejano, and Déborah Oliveros, *Bodies of constant width*, Birkhäuser/Springer, Cham, 2019. An introduction to convex geometry with applications, DOI 10.1007/978-3-030-03868-7. MR3930585

[38] Ernst Meissner and Friedrich Schilling, *Drei Gipsmodelle von Flächen konstanter Breite*, Zeitschrift für angewandte Mathematik und Physik **60** (1912), 92–94.

[39] Alexander Mielke and Michael Ortiz, *A class of minimum principles for characterizing the trajectories and the relaxation of dissipative systems*, ESAIM Control Optim. Calc. Var. **14** (2008), no. 3, 494–516, DOI 10.1051/cocv:2007064. MR2434063

[40] Alexander Mielke and Riccarda Rossi, *Existence and uniqueness results for a class of rate-independent hysteresis problems*, Math. Models Methods Appl. Sci. **17** (2007), no. 1, 81–123, DOI 10.1142/S021820250700184X. MR2290410

[41] Alexander Mielke, Riccarda Rossi, and Giuseppe Savaré, *BV solutions and viscosity approximations of rate-independent systems*, ESAIM Control Optim. Calc. Var. **18** (2012), no. 1, 36–80, DOI 10.1051/cocv/2010054. MR2887927

[42] Alexander Mielke, Riccarda Rossi, and Giuseppe Savaré, *Balanced viscosity (BV) solutions to infinite-dimensional rate-independent systems*, J. Eur. Math. Soc. (JEMS) **18** (2016), no. 9, 2107–2165, DOI 10.4171/JEMS/639. MR3531671

[43] Alexander Mielke and Florian Theil, *On rate-independent hysteresis models*, NoDEA Nonlinear Differential Equations Appl. **11** (2004), no. 2, 151–189, DOI 10.1007/s00030-003-1052-7. MR2210284

[44] Luca Minotti and Giuseppe Savaré, *Viscous corrections of the time incremental minimization scheme and visco-energetic solutions to rate-independent evolution problems*, Arch. Ration. Mech. Anal. **227** (2018), no. 2, 477–543, DOI 10.1007/s00205-017-1165-5. MR3740380

[45] M. Müller, *Konvexe Körper konstanter Breite unter besonderer Berücksichtigung des Meissner-Tetraeders*, Diplomarbeit, Universität zu Köln (2009).

[46] Hans Rademacher and Otto Toeplitz, *The enjoyment of mathematics; Selections from mathematics for the amateur*, Princeton University Press, Princeton, N. J., (1957), MR0081844.

[47] R. Tyrrell Rockafellar and Roger J.-B. Wets, *Variational analysis*, Grundlehren der mathematischen Wissenschaften [Fundamental Principles of Mathematical Sciences], vol. 317, Springer-Verlag, Berlin, 1998, DOI 10.1007/978-3-642-02431-3. MR1491362

[48] Riccarda Rossi and Marita Thomas, *Coupling rate-independent and rate-dependent processes: existence results*, SIAM J. Math. Anal. **49** (2017), no. 2, 1419–1494, DOI 10.1137/15M1051567. MR3636535

[49] Rolf Schneider, *Convex bodies: the Brunn-Minkowski theory*, Encyclopedia of Mathematics and its Applications, vol. 44, Cambridge University Press, Cambridge, 1993, DOI 10.1017/CBO9780511526282. MR1216521

[50] Oded Schramm, *On the volume of sets having constant width [MR968537]*, Selected works of Oded Schramm. Volume 1, 2, Sel. Works Probab. Stat., Springer, New York, 2011, pp. 57–61, DOI 10.1007/978-1-4419-9675-6_3. MR2883369

[51] Giorgio Stefani, *On the monotonicity of perimeter of convex bodies*, J. Convex Anal. **25** (2018), no. 1, 93–102. MR3756927

[52] Augusto Visintin, *Differential models of hysteresis*, Applied Mathematical Sciences, vol. 111, Springer-Verlag, Berlin, 1994, DOI 10.1007/978-3-662-11557-2. MR1329094

[53] Augusto Visintin, *Models of phase transitions*, Progress in Nonlinear Differential Equations and their Applications, vol. 28, Birkhäuser Boston, Inc., Boston, MA, 1996, DOI 10.1007/978-1-4612-4078-5. MR1423808

[54] Kôsaku Yosida, *Functional analysis*, 6th ed., Grundlehren der Mathematischen Wissenschaften [Fundamental Principles of Mathematical Sciences], vol. 123, Springer-Verlag, Berlin-New York, 1980. MR617913

209 South 33rd Street Philadelphia, Pennsylvania 19104-6395
Email address: `rhynd@math.upenn.edu`

Contemporary Mathematics
Volume **781**, 2023
https://doi.org/10.1090/conm/781/15710

A fractional glance to the theory of edge dislocations

Serena Dipierro, Stefania Patrizi, and Enrico Valdinoci

ABSTRACT. We revisit some recents results inspired by the Peierls-Nabarro model on edge dislocations for crystals which rely on the fractional Laplace representation of the corresponding equation. In particular, we discuss results related to heteroclinic, homoclinic and multibump patterns for the atom dislocation function, the large space and time scale of the solutions of the parabolic problem, the dynamics of the dislocation points and the large time asymptotics after possible dislocation collisions.

1. Introduction

1.1. The Peierls-Nabarro model for edge dislocations. The theory of edge dislocations describes defects in the atoms display of a crystal structure. In a nutshell, dislocations represent areas of a crystal in which the atoms do not lie in the geometrically organized pattern of the material, and they are responsible for plastic deformations to occur. In particular, these plastic deformations occur without altering the chemical composition of the material and under external forces significantly lower than expected.

As an example of dislocation, one can think of a simple cubic lattice representing the crystal at a large scale and suppose that an irregularity arises in the lattice: see for instance Figure 1, in which an imperfection of the crystal is visible on the fourth horizontal row.

As a matter of fact, Figure 1 must be understood as (the local reproduction of) a planar section of a three-dimensional (infinitely extended) crystal and represents the "typical" anomaly associated to an edge dislocation, as being caused by the termination of a plane of atoms in the middle of a crystal (the picture is similar to that of sticking half of a piece of paper into a stack of paper; in this case the half of a piece of paper corresponding to the three extra atoms visible in the upper part of the third column of atoms in Figure 1).

A pioneer description of elastic dislocations dates back to Vito Volterra [**Vol07**] and the connection between plastic deformations and dislocation theory was set forth by Egon Orowan, Michael Polanyi and Geoffrey Taylor (independently and at about the same time) explaining that the stress needed for the slip was far lower than theoretical predictions in view of the theory of dislocations [**Oro34a**, **Oro34b**, **Oro34c**, **Pol34**, **Tay34a**, **Tay34b**]. An intuitive reason for why dislocation dynamics allow plastic displacements under a low external force can be found by comparing

2020 *Mathematics Subject Classification.* Primary .

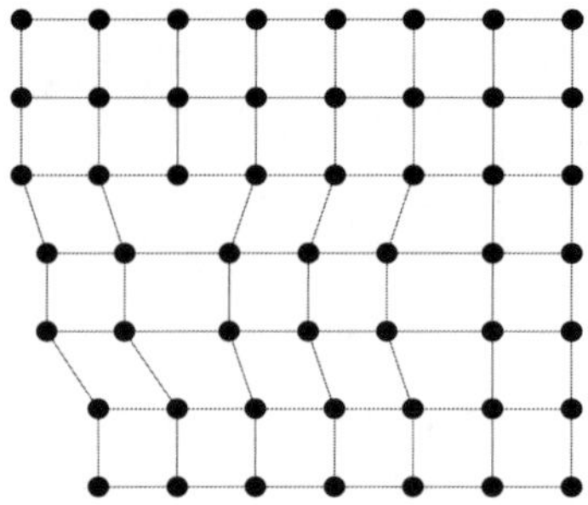

FIGURE 1. A defect in a crystal (the black dots corresponding to atoms and the cyan segments to the bonds between atoms).

this phenomenon with the amusing locomotion of caterpillars obtained by squeezing muscles in sequence in an undulating wave motion: namely, since caterpillars do not have any bone in their entire bodies, they can produce a "defect" by arching their body and obtain a displacement by moving this defect along their body (this is however a rather crude biological explanation, if interested in a detailed analysis of the caterpillars' locomotion see e.g. [**vGT14**]).

At that time, the description of dislocations was mainly based on theoretical predictions with little or no experimental evidence for dislocations. Indeed, although slip lines on the surface of metals had been observed by optical microscopy, a direct observation of crystal dislocations arrived later, since it required transmission electron microscopes [**Ver51, Ame52, Bol56, HHW56, SH59, HCSW06**].

A useful graphical device to detect crystal defects in a lattice is given by the so-called Burgers vector, named after Johannes (Jan) Martinus Burgers [**Bur40**], see Figure 2. The gist is that in a perfect crystal structure if one moves for a given number of steps, say three, in every coordinate direction, say right-down-left-up, the trajectory produced is a square that returns to the initial point; instead, when a similar ideal movement is drawn encompassing a dislocation, the arrival point is different from the initial one and the vector joining the starting position with the final one is a vector, namely the Burgers vector, which encodes the magnitude and direction of the crystal imperfection.

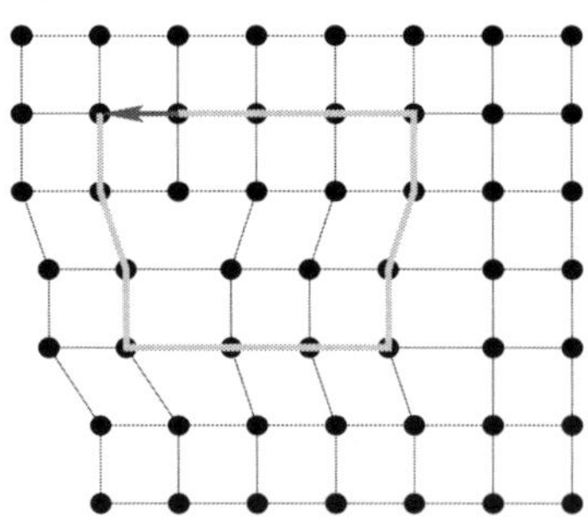

FIGURE 2. Burgers vector.

From the static viewpoint, a crystal defect produces a structural inhomogeneity corresponding e.g. in Figure 1 to a relative compression above the dislocation, a relative tension below and a shear on the side (the shear can be visualized since the bonds between atoms are shaped in the form of a parallelogram rather than a regular square).

In practice, one can take advantage of this structural imperfection by producing plastic deformations as an outcome of the motion of dislocations which is induced by a set of shearing forces, namely forces parallel to one of the axis of the lattice, pushing the crystal in one specific direction from a side, and in the opposite direction from the other side (whatever "side" may mean here, since our ideal crystal has infinite extension in every direction). See Figure 3 for a description of shearing forces.

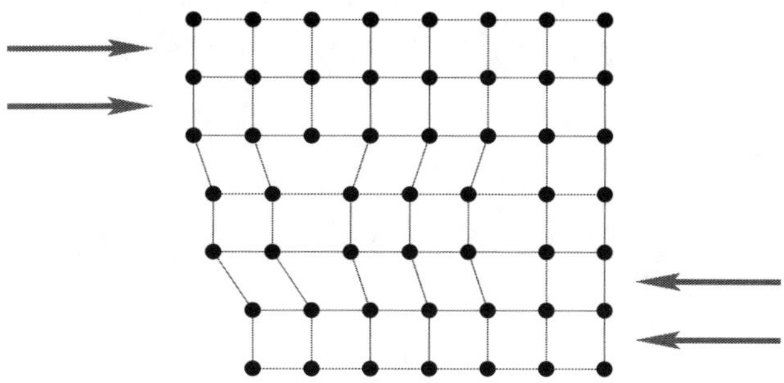

FIGURE 3. Shearing forces.

The advantage of dislocations towards plastic deformations is that when the shearing forces are parallel to the Burgers vector their result can be that of weakening or breaking some of the bonds between atoms and allow them to slide over each other. This glide, or slip, method is thus capable of producing plastic deformations at low stress levels. See Figure 4 for a cartoon of how a shearing force can move around dislocations in the lattice with the final outcome of producing a visible plastic deformation: the idea is that by conveniently sliding a dislocation the crystalline order is restored on either side but the atoms on one side have moved by one position.

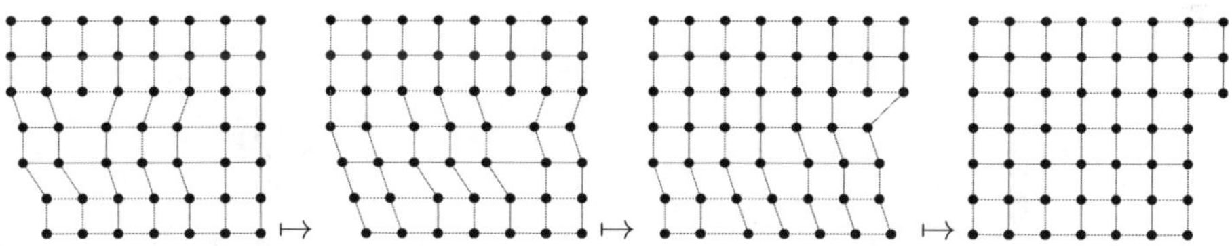

FIGURE 4. Plastic deformation as an outcome of crystal dislocation motion.

We have a direct experience of the role played by dislocations in plastic deformations anytime we indulge in playing around with paperclips (those little devices made of steel wire bent to a looped shape that we use to hold sheets of paper together before every useful information was digitalized and posted on the internet, see Figure 5): with soft movements of our fingers, we can easily open and close repeatedly the looped shape of a paperclip, enjoying its flexibility and malleability – till at some point the paperclip becomes suddenly quite stiff and to change its shape we need to exert some extra force which typically ends up breaking the paperclip. Hence, this simple experiment somehow confirms how the strength and ductility of metals are influenced and controlled by dislocations.

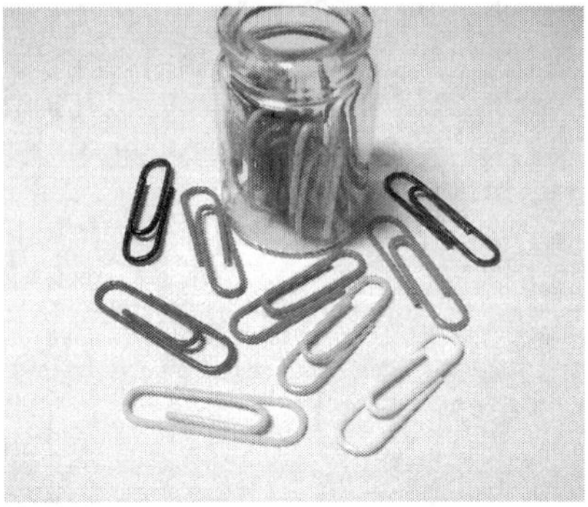

FIGURE 5. A few paper clips of different colors (Public Domain image from Wikipedia).

The physical explanation of this phenomenon is that at the beginning the paperclip structure presents some defects and our soft movements rely on the dynamics of these dislocations to modify the shape of the object; at some point, however, several of these imperfections get resolved precisely by our repeated opening and closing the loop of the clip, and when no more dislocation is available (or when too few dislocations are available) the perfect (or almost perfect) lattice offers a sounder resistance to the external stress (see the last frame in Figure 4), which results in the (perhaps not so expected) stiffness that typically make this game end by the breaking of the clip.

Some technical remarks are in order. First off, the assumption that the crystal has a simple cubic structure (also called primitive cubic in jargon) is indeed a great geometric simplification but it includes several concrete cases of interest, such as polonium Po, see [**Sch07**], iron pyrite FeS_2, see Figure 6, as well as an allotrope of selenium Se, see [**Sha07**].

Furthermore, we stress that we are focused here on edge dislocations, namely we only consider here situations in which dislocations move in the direction of the Burgers vector (as in Figures 3, 4 and 7). Other situations also occur in nature, such as the so-called screw dislocations in which the dislocation dynamics takes place in a direction perpendicular to the Burgers vector, thus creating a helical arrangement of atoms around the core, see e.g. Figure 8. Mixed dislocations exist as well, in which case the Burgers vector is neither parallel nor orthogonal to the dislocation dynamics. But in this note we confine our interest to the simplest case of edge dislocations.

To mathematically describe edge dislocations, we recall the classical model by Rudolf Peierls and Frank Nabarro. Suppose that the crystal is subject to shearing forces causing a slip along a horizontal plane, that we call the slip plane. We take a vertical section of the crystal, see Figure 7. The intersection between the horizontal slip plane and the vertical sectional plane is called slip line. We suppose that at a large scale the crystal has a simple cubic structure described by the lattice $\mathbb{Z}^3$ and

FIGURE 6. Pyrite cubic crystals (image from Wikipedia, licensed under the Creative Commons Attribution-Share Alike 3.0 Unported license).

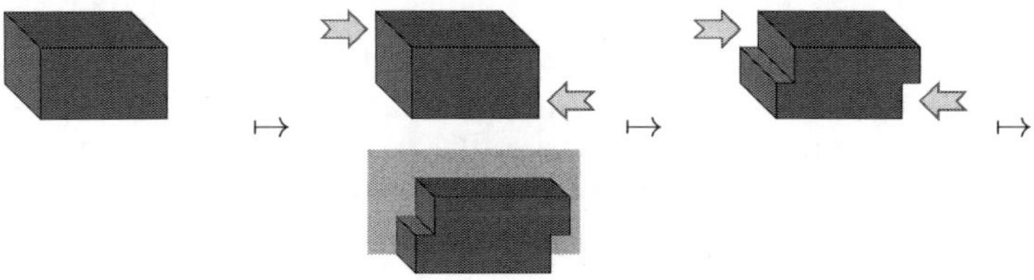

FIGURE 7. Slip of a crystal along a horizontal plane in an edge dislocation, and vertical section.

accordingly the vertical sectional plane is such that it detects a large scale atom disposition described by the lattice $\mathbb{Z}^2$.

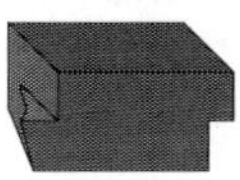

FIGURE 8. A screw dislocation.

We stress that this lattice does not describe the position of every single atom, since defects at small scales can occur. The dislocation theory thus makes the ansatz that the lattice structure simply dictates a favorable state for atoms which are however free to move, subject to a potential that penalizes the atom positions outside the lattice structure and a bond interaction between atoms.[1]

[1]To favor the intuition, one can think about people in a room with chairs placed in a lattice structure. Suppose that people have some bonds between them (either psychological, being influenced by the movements of the people around and trying to follow their neighbors, or physical, e.g. holding hands). Suppose also that people have a preference to sit comfortably on a chair (not necessarily "their" chair, any chair is equally good). Most chairs are occupied, but some chair is free; most people sit on a chair, but perhaps someone could not find a convenient place. Then, if

The slip will indeed cause some misalignment between the atoms below the slip lines and the ones above. For concreteness, we focus our attention on the atoms above the slip line (the case of the ones below being similar) and we aim at describing their mismatch with respect to the lattice structure.

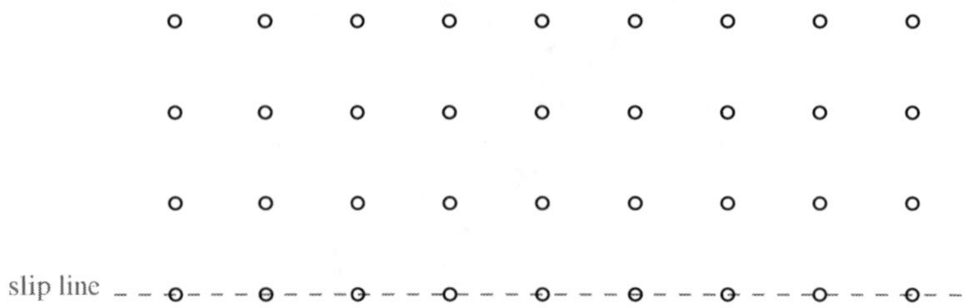

FIGURE 9. Rest position of the atoms corresponding to the large scale crystalline structure.

For this, we consider the atom rest positions $\mathbb{Z}^2$ in the upper halfplane $\mathbb{R} \times [0, +\infty)$, corresponding to the empty circles in Figure 9.

Positions in this halfplane are denoted by using coordinates $(x, y) \in \mathbb{R} \times [0, +\infty)$. The position of each atom is described by a dislocation function that identifies the mismatch with respect to the rest position: for simplicity, we assume that atoms can only slide horizontally (that is, parallel to the slip plane and to the shearing forces), see Figure 10 in which the full circles represent the atoms (we are thus neglecting the vertical mismatch of atoms possibly caused by mutual bonds, by considering this effect as negligible with respect to the shearing force).

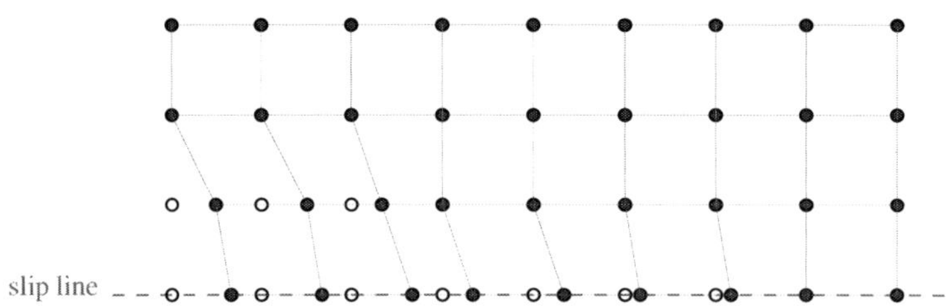

FIGURE 10. Mismatch between atom location and crystal structure.

In this setting, the distance between the actual position of an atom and its rest position located at $(x, y) \in \mathbb{R} \times [0, +\infty)$ corresponds to a scalar function $U(x, y)$ (say, with U positive if the atom is dislocated towards the right and negative if the atom is dislocated towards the left, see Figure 11). It is also convenient to consider the trace of the dislocation function U along the slip line which corresponds to $u(x) := U(x, 0)$.

---

some of the people move around, they may drag with them others, but the collective interest is to find a chair to sit down.

With respect to the crystal dislocation model, people correspond to the atoms, the chair to the atom favorable position aligned with the cubic lattice, the empty chairs to the crystal defects. A quantitative difference is that typically the potential describing people's satisfaction is discontinuous (one would charge zero discomfort if someone sits on a chair and unit discomfort if not, no matter how far or how close the individual is to the closest chair), while the potential used in physical applications are typically of multi-well shapes (the minima corresponding to the optimal location on the lattice), but continuous and in fact smooth.

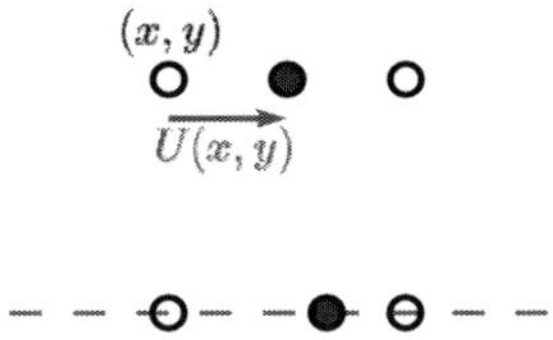

FIGURE 11. The dislocation function $U(x, y)$.

In a situation such as the one described in Figure 7, we can argue that most of the mismatch occurs along the slip line: roughly speaking, we can expect the atoms located in the upper halfplane to be mostly alligned with a cubic lattice and the ones located in the lower halfplane to be mostly alligned with another cubic lattice which is a horizontal translation of the lattice located in the upper halfplane. Therefore, we can suppose that the main contribution to the crystal mismatch is encoded[2] by the function u. To quantify this mismatch, we consider a multiwell potential W, say a scalar function such that $W(k) = 0$ for all $k \in \mathbb{Z}$ and $W(r) > 0$ for all $r \in \mathbb{R} \setminus \mathbb{Z}$. For simplicity, we will suppose in what follows that W is smooth, periodic of period 1, even and[3] such that

$$W''(0) \neq 0. \tag{1.1}$$

This potential penalizes atoms which are not aligned with the lattice structure, since $W(u(x))$ turns out to be positive unless $u(x) \in \mathbb{Z}$, that is unless the atom with ideal rest position at x is located to a point of the lattice (recall that any point of the lattice ends up to be good for the atom to be in equilibrium with respect to the large scale structure of the crystal).

As a result, in this model the total energy contribution coming from atom displacement can be written as

$$\mathscr{P}(U) := \int_{\mathbb{R}} W(u(x))\, dx.$$

The total energy however has to account also for the bonds between atoms. The simplest ansatz is to suppose that this energy contribution resembles an elastic

[2]As stressed after (1.3), it is now convenient to think that the rest positions of the atoms correspond to a *discrete lattice*, but the matter itself is a *continuum*. This is perhaps conceptually a bit confusing, since one ends up considering the dislocation function as a (smooth) object defined on a continuum, which identifies a sort of mismatch of the state of matter with respect to a discrete crystal structure: roughly speaking, given any point in a continuous substance, the dislocation function determines how far a state parameter is from that of a perfect lattice. In practice, this confusion between discrete and continuous objects may turn out to be acceptable, given the very small scale at which the crystal structure becomes visible. As a result, one is empowered to use the robust tools of continuous mathematical analysis in a problem which seems discrete in nature.

[3]Though widely exploited in the literature, condition (1.1) is not merely technical, since the nondegeneracy of the minima of the potential produces a linearized equation with nontrivial linear forcing term, which in turn has an impact on the decay of the solutions. The counterpart of this in the analysis of the Allen-Cahn equation is that condition (1.1) is not needed for the theory of Γ-convergence, but it does play a role to establish density estimates and uniform convergence of level sets. For the Allen-Cahn equation, some degenerate cases can also be thoroughly treated, see [**DFV18**], and in further works we plan to extend the theory of crystal dislocations also to singular and degenerate potentials. Other interesting cases to deal with take into account situations in which a singular or degenerate potential is confronted to a singular or degenerate elastic term (see [**Val06, VSS06**] for the counterpart of these cases for the Allen-Cahn equation).

energy of the form

$$\mathscr{E}(U) := \frac{1}{2} \int_{\mathbb{R}\times(0,+\infty)} |\nabla U(x,y)|^2 \, dx \, dy, \tag{1.2}$$

where the factor $\frac{1}{2}$ has been introduced for later convenience (we are disregarding here structural constants).

The total energy is therefore the sum of the displacement energy and the bond energy and takes the form

$$\mathscr{H}(U) := \mathscr{E}(U) + \mathscr{P}(U) = \frac{1}{2} \int_{\mathbb{R}\times(0,+\infty)} |\nabla U(x,y)|^2 \, dx \, dy + \int_{\mathbb{R}} W(u(x)) \, dx. \tag{1.3}$$

We stress that this model is, in a sense, of "hybrid" type: while the rest position of the atoms correspond to a lattice structure with a discrete feature, the dislocation function U (and thus its trace u) is assumed to be continuous (and actually smooth) and defined in the continuum given by $\mathbb{R} \times (0, +\infty)$. This "philosophical inconsistency" between the discrete and continuous descriptions of nature is compensated by a simple and effective realization of the model and it is also, at least partially, justified by the fact that the atom distance is typically much smaller than the plastic effects that one is interested in taking into account (therefore, for practical purposes, the lattice acting as a substratum is "almost" a continuum).

Equilibrium configurations of this model correspond to critical points of $\mathscr{H}$. These configurations can be detected by noticing that, for every $R > 0$ and any smooth function $\Psi : \mathbb{R} \times [0, +\infty) \to \mathbb{R}$ such that $\Psi(x, y) = 0$ if $|(x, y)| \geqslant R$,

(1.4)

$$\begin{aligned} 0 &= \langle D\mathscr{H}(U), \Psi \rangle \\ &= \int_{\mathbb{R}\times(0,+\infty)} \nabla U(x,y) \cdot \nabla \Psi(x,y) \, dx \, dy + \int_{\mathbb{R}} W'(u(x))\psi(x) \, dx \\ &= - \int_{\mathbb{R}\times(0,+\infty)} \Delta U(x,y)\Psi(x,y) \, dx \, dy + \int_{\mathbb{R}} \big(W'(u(x)) - \partial_y U(x,0)\big)\psi(x) \, dx, \end{aligned}$$

where $\psi(x) := \Psi(x, 0)$, thanks to Green's first identity.

In particular, if we take a ball $B \Subset \mathbb{R} \times (0, +\infty)$ and a function $\Psi_0 \in C_0^\infty(B)$, it follows that $\psi_0(x) := \Psi_0(x, 0) = 0$ for all $x \in \mathbb{R}$ and thus, by (1.4),

$$0 = - \int_{\mathbb{R}\times(0,+\infty)} \Delta U(x,y)\Psi_0(x,y) \, dx \, dy,$$

that leads to U being harmonic in $\mathbb{R} \times (0, +\infty)$.

Inserting this information into (1.4), we find that

$$0 = \int_{\mathbb{R}} \big(W'(u(x)) - \partial_y U(x,0)\big)\psi(x) \, dx,$$

giving that $W'(u) = \partial_y U$ along the slip line.

From these considerations, we infer that equilibria of the Peierls-Nabarro model correspond to solutions of

$$\begin{cases} \Delta U = 0 & \text{in } \mathbb{R} \times (0, +\infty), \\ \partial_y U = W'(u) & \text{in } \mathbb{R} \times \{0\}. \end{cases} \tag{1.5}$$

Formally, using the harmonicity of U and the Poisson kernel for the halfspace, we thus obtain that, for all $(x,y) \in \mathbb{R} \times (0,+\infty)$,

$$U(x,y) = \frac{1}{\pi} \int_{\mathbb{R}} \frac{y\, u(\zeta)}{(x-\zeta)^2 + y^2}\, d\zeta.$$

To remove the singularity as $y \searrow 0$, one can notice that a primitive of $\mathbb{R} \ni \zeta \mapsto \frac{y}{(x-\zeta)^2+y^2}$ is $\arctan\left(\frac{\zeta - x}{y}\right)$ and accordingly

$$\frac{1}{\pi} \int_{\mathbb{R}} \frac{y}{(x-\zeta)^2 + y^2}\, d\zeta = \frac{1}{\pi}\big(\arctan(+\infty) - \arctan(-\infty)\big) = 1.$$

This gives that

$$U(x,y) - u(x) = \frac{1}{\pi} \int_{\mathbb{R}} \frac{y\,\big(u(\zeta) - u(x)\big)}{(x-\zeta)^2 + y^2}\, d\zeta$$

and therefore

$$\begin{aligned} W'(u(x)) &= \lim_{y \searrow 0} \partial_y U(x,y) = \lim_{y \searrow 0} \partial_y \big(U(x,y) - u(x)\big) = \frac{1}{\pi} \lim_{y \searrow 0} \int_{\mathbb{R}} \frac{\partial}{\partial y} \left(\frac{y\,\big(u(\zeta) - u(x)\big)}{(x-\zeta)^2 + y^2} \right) d\zeta \\ &= \frac{1}{\pi} \lim_{y \searrow 0} \int_{\mathbb{R}} \frac{(\zeta^2 - 2x\zeta + x^2 - y^2)\,\big(u(\zeta) - u(x)\big)}{(\zeta^2 - 2x\zeta + x^2 + y^2)^2}\, d\zeta = \frac{1}{\pi} \int_{\mathbb{R}} \frac{u(\zeta) - u(x)}{\zeta^2 - 2x\zeta + x^2}\, d\zeta \\ &= \frac{1}{\pi} \int_{\mathbb{R}} \frac{u(\zeta) - u(x)}{(x-\zeta)^2}\, d\zeta, \end{aligned}$$

that is, neglecting normalizing constants,

$$-\sqrt{-\Delta}\, u = W'(u) \quad \text{in } \mathbb{R}. \tag{1.6}$$

Alternatively, one can formally obtain (1.6) from (1.5) from the analysis of a Dirichlet-to-Neumann problem (see e.g. [**BV16**, Section 4.1]) or solving the linear equation in (1.5) via Fourier transform (see e.g. [**BV16**, Section 4.3]).

We think that (1.6) is a neat example of a nonlocal equation of fractional type arising from purely classical considerations. Of course, equation (1.6) presents both advantages and disadvantages with respect to (1.5). On the one hand, equation (1.5) is classical in nature and it may seem more handy to be dealt with. On the other hand, equation (1.6) showcases the intrinsic nonlocal aspect of the problem, in which dragging one single atom along the slip line ends up influencing, in principle, the whole crystalline structure. Furthermore, equation (1.6) is set in dimension 1, which is a considerable conceptual advantage since it opens the possibility of employing methods from dynamical systems (which will provide essential ingredients in Section 2).

Additionally, equation (1.6) immediately allows to consider more general equations of the form

$$-(-\Delta)^s u = W'(u) \quad \text{in } \mathbb{R}, \tag{1.7}$$

for a given $s \in (0,1)$.

In this spirit, equation (1.7) is not a mere generalization of (1.6) motivated by mathematical curiosity: instead, it can offer a useful approach to comprise models in which the range-effect of the atomic bonds is different from the one modeled by the purely elastic response introduced in (1.2). Moreover, it allows a direct comparison between equations arising in material sciences and crystal dislocation dynamics with similar ones motivated, for instance, by problems in ethology, population dynamics and mathematical biology (for instance, similar models can be considered to describe a biological population subject to a nonlocal dispersal strategy of Lévy flight type, see e.g. [**DGV22**] and the references therein).

Also, and possibly more importantly, considering (1.6) as a "particular case" of (1.7) allows one to take into account a continuous range of fractional exponents s, avoiding "discrete jumps" from the classical case $s = 1$ (often related to the classical Allen-Cahn equation) and the case $s = \frac{1}{2}$ in (1.7): besides its conceptual advantage, this approach allows one to exploit perturbative methods in the fractional exponent (e.g. bifurcating from previously known cases).

1.2. Organization of the paper. The rest of the article is organized as follows. In Section 2 we present some recent results that describe situations in which atom dislocations in crystals can display complicated or unusual patterns as stationary equilibria, thus producing chaotic behaviors.

In Section 3 we analyze the dislocation dynamics and in Section 4 the cases in which collisions between dislocation points occur. Section 5 is then devoted to discussing what happens to the dislocations after collisions.

In Section 6 we deduce the so-called Orowan's Law, and in Section 7 we deal with the homogenization problem associated with the Peierls-Nabarro model.

2. Chaotic dislocation displacements

A natural question is whether atom dislocations in crystals can display complicated or unusual patterns as stationary equilibria. We started the investigation of this question in [**DPV17, DPV19**]. The gist of this analysis is that, without indulging in philosophical definitions, one of the simplest forms of "chaos" consists in the possibility of producing patterns which exhibit transition layers and oscillatory behaviors. To accomplish this goal, we find it convenient to revisit some classical methods from dynamical systems and topological analysis (such as the ones in [**CZR91, RCZ00**]) in light of the fractional setting introduced in (1.7).

From the technical point of view, we recall that the existing methods were designed for classical Hamiltonian or Lagrangian systems and were aiming at detecting oscillatory behaviors in the time variable, that is for the time evolution of a particle, or a system of particles. The structure in (1.7) is different, since no time variable is involved: instead, one can use methods from dynamical systems to detect oscillatory phenomena in the space variable, especially thanks to the fact that equation (1.7) is set in one space variable (after all, once things are set into a good mathematical formulation, one space variable can formally play the role of time, and vice versa).

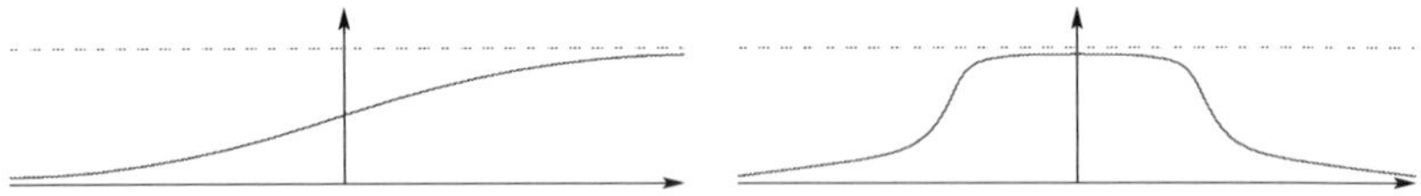

FIGURE 12. A heteroclinic and a homoclinic connections.

The goal in [**DPV17, DPV19**] is thus to construct complex examples of stationary equilibria for the atom dislocation function induced by a perturbation of the potential (roughly speaking, pressing or pulling a bit the crystal). The complex example that we were interested in included:

- heteroclinic connections, i.e. transition layers connecting two minima of the potential W,
- homoclinic connections, i.e. a nontrivial connection of a minimum of the potential W to itself,
- multibump orbits, i.e. solutions oscillating somewhat arbitrarily between the minima of the potential W,

see[4] Figures 12 and 13.

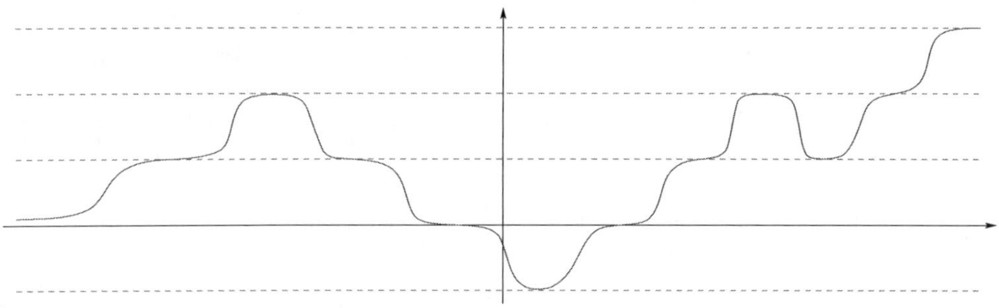

FIGURE 13. A multibump orbit.

To give a consistent mathematical framework to this perturbation setting, instead of (1.7) one considers the equation

$$(-\Delta)^s u(x) + a(x)\, W'(u(x)) = 0 \qquad \text{for any} \quad x \in \mathbb{R}. \tag{2.1}$$

The role of a is that of a "generic" (arbitrarily small and with small derivatives, but nontrivial) perturbation: to avoid technicalities related to the precise nondegeneracy assumptions on a, one can just stick to the case in which

$$a(x) := 1 + \delta\big(1 - \cos(\eta x)\big), \tag{2.2}$$

where δ and η are given, positive, small parameters (note that when either $\eta = 0$ or $\delta = 0$ the function a reduces to the constant 1, hence a can be thought as an arbitrarily small perturbation with arbitrarily small frequency).

Among other results, it is proved in [**DPV17**] that

THEOREM 1. *Let $s \in \left(\frac{1}{2}, 1\right)$, $N \in \mathbb{N}$, $\zeta_1, \ldots, \zeta_N \in \mathbb{Z}$, with $|\zeta_{i+1} - \zeta_i| = 1$, for $i \in \{1, \ldots, N-1\}$.*

Then, there exist $b_1 < \cdots < b_{2N-2} \in \mathbb{R}$ and a solution u of (2.1) such that:

$$\lim_{x \to -\infty} u(x) = \zeta_1,$$

$$\sup_{x \in (-\infty, b_1]} |u(x) - \zeta_1| \leqslant \frac{1}{100},$$

$$\sup_{x \in [b_{2i}, b_{2i+1}]} |u(x) - \zeta_{i+1}| \leqslant \frac{1}{100} \qquad \textit{for all } i \in \{1, \ldots, N-2\},$$

$$\sup_{x \in [b_{2N-2}, +\infty)} |u(x) - \zeta_N| \leqslant \frac{1}{100}$$

and $\displaystyle\lim_{x \to +\infty} u(x) = \zeta_N.$

---

[4]We stress that Figures 12 and 13 refer to possible trace functions u. As such, they are depicted as functions from $\mathbb{R}$ to $\mathbb{R}$.

In this spirit, Theorem 1 constructs a spatially oscillatory equilibrium for the dislocation function as the one depicted in Figure 13: we stress that the levels $\zeta_1, \dots, \zeta_N \in \mathbb{Z}$ can be prescribed arbitrarily as consecutive integers and therefore Theorem 1 states that a generic perturbation can produce dislocation patterns that oscillate essentially as one wishes; we recall that, in view of the notation stated on page 108, a positive oscillation of the dislocation function would correspond to an atom dislocation "right" with respect to its rest position, and a negative oscillation of the dislocation function would correspond to an atom dislocation "left" with respect to its rest position and accordingly the oscillations "up and down" detected in Theorem 1 would correspond to atoms arbitrarily dislocated to the left and to the right of the lattice in the setting of Section 1.1. Being able to dislocate atoms at will opens the possibility of associating to the atomic structure a "symbolic dynamics" that describes a state of the system in terms of the corresponding minimum of the potential W.

Particular cases of Theorem 1 include the situations displayed in Figure 12, namely:

- when $N := 2$ and $\zeta_2 := \zeta_1 + 1$ (or $N := 2$ and $\zeta_2 := \zeta_1 - 1$), one obtains a heteroclinic orbit,
- when $N := 3$ and $\zeta_3 := \zeta_1$, one obtains a homoclinic orbit.

We observe that while heteroclinic orbits also exist in case of a constant perturbation a, the existence of homoclinic and multibump orbits heavily depends on the fact that a is supposed to be generic (and in particular nonconstant): indeed, heteroclinic orbits for constant a have been constructed in [**PSV13**, **CS15**] but it has been shown in [**CDdP20**] that no homoclinic connection and no multibump solution are possible when a is constant.

The result in Theorem 1 is actually flexible enough to deal also with more general interaction kernels and with equilibria not necessarily disposed on a regular lattice. The result also applies to systems (though in this case the subsequent layer ζ_{i+1} is not completely arbitrary given ζ_i, since it must be chosen as a suitable action minimizers among close neighbors).

In terms of methodology, while Theorem 1 takes its inspiration from classical variational methods, constrained minimization techniques and energy estimates, its proof needs to account for some of the difficulties provided by the nonlocal setting of equation (2.1). In particular:

- the "cut and paste methods" used in dynamical systems to create competitors for the energy functional and obtain energy estimates are affected in our case by nonlocality (for instance, the energy of the union of two trajectories is not the sum of their energies),
- the long tails of the solutions provide significant energy contributions (suggestively, as in what Einstein called Mach's principle, "mass out there influences inertia here" and "local physical laws are determined by the large-scale structure of the universe" [**HE73**]).

To prove Theorem 1, the variational method relies on a constrained minimization problem: roughly speaking, one first minimizes the energy functional under the additional request that the orbit passes through suitable "windows" strategically located to create the desired oscillations (namely, the cyan boxes in Figure 14).

However, to make this argument work, one needs to avoid that the minimizing orbit constructed in this way touches the boundary of the window (indeed, if it

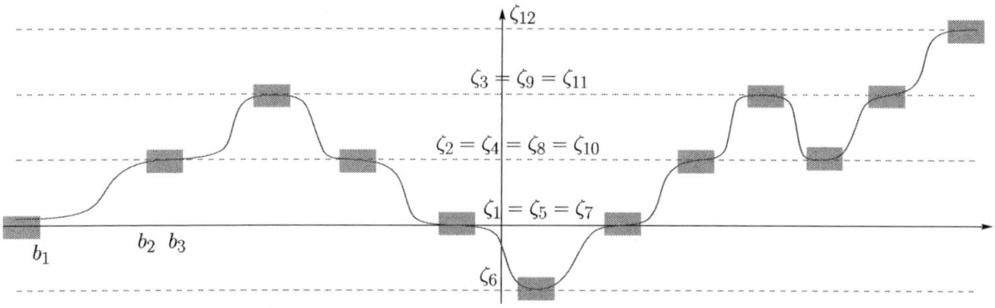

FIGURE 14. Constrained minimization to construct a multibump orbit.

does not, then small perturbations around the orbits are admissible and we obtain a local minimizer, hence a solution of the problem with the desired oscillations).

To get rid of the possible touching between the minimal orbit and the windows, one needs to carefully choose the intervals with endpoints $b_1, \dots, b_{2N-2}$ and exploit the oscillations of a to place the transitions in a favorable energy range.

To clarify this method, let us focus on the case of heteroclinics (which will also serve as a cornerstone to build the other kinds of trajectories), namely let us deal with the case $N := 2$ and $\zeta_2 := \zeta_1 + 1$. Suppose that the method of constrained minimization "goes wrong", that is a touching occurs as shown by the magenta circle in Figure 15.

One first shows that indeed the touching point must occur quite close to the boundary of the window (at infinity, it is energetically more convenient for the solution to stay close to the minima of the potential and thus away from the window). Then, one needs to show that the transition from ζ_1 to ζ_2 occurs in a rather short interval (it is energetically inconvenient for the solution to stay away from the minima of the potential in a large interval). Finally, once the solution gets sufficiently close to ζ_2 on the right, it is better for it to remain close to it (it is energetically inconvenient for the solution to drift away from the minima of the potential once it gets there).

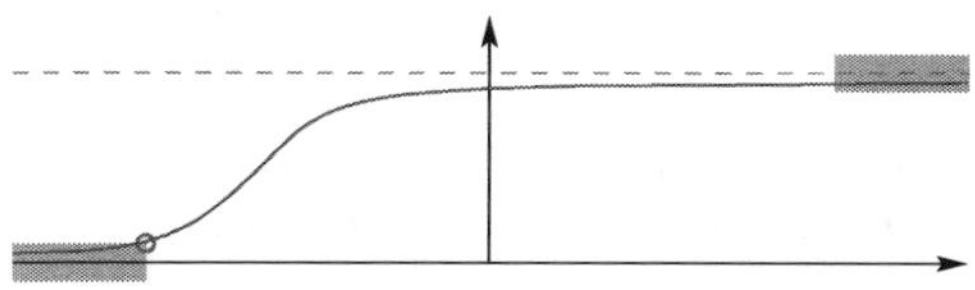

FIGURE 15. The constrained minimization method going wrong.

With these three ingredients in mind, we realize that indeed Figure 15 represents essentially the "worst case scenario" that we want to avoid. For this, the idea is to translate the red trajectory in Figure 15 to the right and lower its energy: to accomplish this, one needs to assume that the modulating perturbation a presents some minima that we can enclose between the two cyan windows (hence, to this end, if a is slowly oscillating one needs to place the windows sufficiently far from each other). Since the effect of a is to modulate the potential W and since we expect "most of the energy" to come from the transition from ζ_1 to ζ_2, it will be energetically more convenient to locate the transition inside the minima of a, thus obtaining a new minimizer that does not touch the window, as desired (see Figure 16).

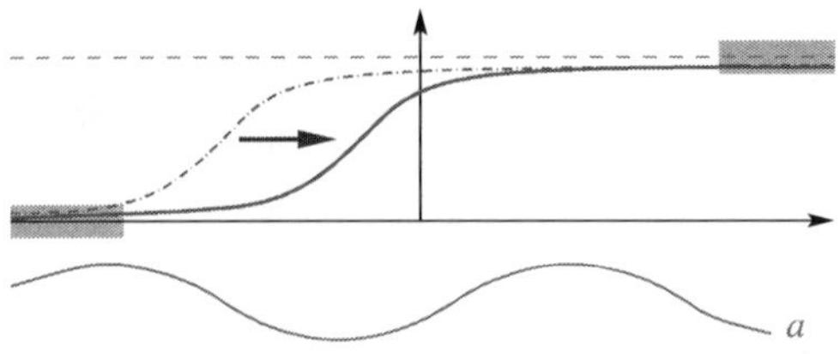

FIGURE 16. Translating the orbit into an energetically more favorable position.

When implementing these ideas, however, some effort is needed to take care of nonlocality, since energy competitors cannot be obtained simply by gluing pieces of trajectories (this feature will also become pivotal when one wants to address Theorem 1 in its full generality, since homoclinic and multibump solutions will be obtained precisely by "almost" glueing heteroclinics together). To minimize the impact of nonlocality on the energy computations, we glue orbits at "very flat" matching points, namely at points at which two orbits meet with very small derivative, remaining close to each other in a large interval (we quantify appropriately this notion and we call such intervals "clean intervals"). The existence of these clean intervals is obtained by combining energy estimates and fractional elliptic regularity theory.

We observe that Theorem 1 was obtained in the fractional exponent range $s \in \left(\frac{1}{2}, 1\right)$ which is arbitrarily close to, but different from, the exponent $s = \frac{1}{2}$ arising in (1.6). The case $s = \frac{1}{2}$, or more generally $s \in \left(0, \frac{1}{2}\right]$, is indeed quite challenging, due to several additional difficulties. First of all, when $s = \frac{1}{2}$ one has to account for unbounded competitors, namely functions with finite energy and unbounded spikes (which would make the constrained minimization problem with pointwise windows unpractical). To appreciate this hindrance, one may consider the function

$$\bar{\psi}(x) := \chi_{(-1,1)}(x) \ln(1 - \ln|x|)$$

which belongs to $H^{1/2}(\mathbb{R}) \setminus L^\infty(\mathbb{R})$, see [**DPV19**, Appendix A]. From this, one can easily construct competitors with small energy but unbounded spikes, e.g. both accumulating at some given points and at infinity, such as

$$\bar{\psi}_\varepsilon(x) := \varepsilon \sum_{k \in \mathbb{Z}} \frac{\bar{\psi}(2^{|k|}(x - 2^k))}{2^{|k|}},$$

see Figure 17.

An additional difficulty comes from the fact that when $s = \frac{1}{2}$ heteroclinic solutions have infinite energy: to appreciate[5] this, one can argue that, roughly speaking, if

(2.3)
$$u_\star \text{ solves (1.7) and is monotone and such that } u_\star(-\infty) = 0 \text{ and } u_\star(+\infty) = 1,$$

[5]When $s = \frac{1}{2}$ and $W(r) = \cos r$, explicit heteroclinics are known, see e.g. [**AV19**, Appendices L and M]. See also [**AV19**, Appendix K] for more rigorous decay estimates than the ones sketched here.

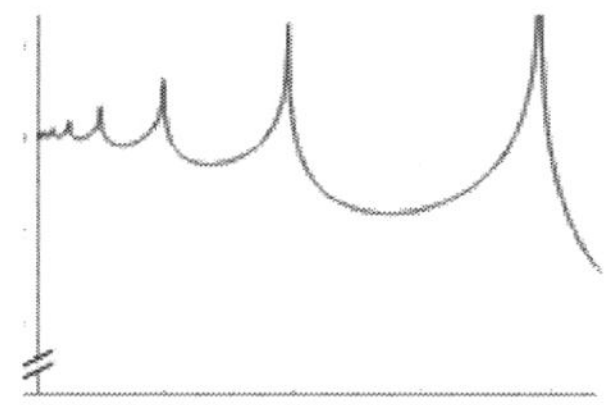

FIGURE 17. A small energy competitor with many unbounded spikes.

then, setting $w := u'_\star$, at $+\infty$ one obtains a behavior of the type

$$-(-\Delta)^s w(x) = W''(u_\star(x))w(x) \simeq W''(1)w(x),$$

that is, for large x,

$$w(x) \simeq \int_{\mathbb{R}} \frac{w(\zeta) - w(x)}{|x-\zeta|^{1+2s}}\, d\zeta \simeq \int_{\mathbb{R}} \frac{w(\zeta)}{|x-\zeta|^{1+2s}}\, d\zeta \simeq \frac{C}{x^{1+2s}}$$

and consequently

$$u_\star(x) \simeq 1 - \frac{C}{x^{2s}}. \tag{2.4}$$

This shows that when $s = \frac{1}{2}$, and more generally when $s \in \left(0, \frac{1}{2}\right]$, the heteroclinics possess[6] infinite elastic energy, because, for large R,

$$\begin{aligned}\iint_{\mathbb{R}\times\mathbb{R}} \frac{|u_\star(x) - u_\star(y)|^2}{|x-y|^{1+2s}}\, dx\, dy &\geqslant \frac{1}{4} \iint_{(R,+\infty)\times(-\infty,-R)} \frac{dx\, dy}{(x-y)^{1+2s}} \\ &= \frac{1}{8s} \int_{(R,+\infty)} \frac{1}{(x+R)^{2s}}\, dx = +\infty.\end{aligned}$$

These observations clarify that for $s = \frac{1}{2}$, and more generally for $s \in \left(0, \frac{1}{2}\right]$, the standard variational methods need to be modified since the objects that we seek have infinite energy while the objects that we want to avoid (such as the ones presenting uncontrolled spikes) can have finite, and even small, energy.

These difficulties have been addressed in [**DPV19**], where heteroclinic connections for equation (2.1) have been constructed in the fractional exponent range $s \in \left(\frac{1}{4}, \frac{1}{2}\right]$. To overcome the difficulty arising from the infinite energy of the heteroclinics, we use "energy renormalization" methods (by formally subtracting the infinite elastic energy of the heteroclinic to the functional). Also, to avoid finite energy competitors which may be discontinuous, and even unbounded, we employ an "energy penalization" method, by adding a "viscosity" term to the equation,

[6]It is instructive to remark that instead the potential energy is finite when $s \in \left(\frac{1}{4}, 1\right)$ since in this range

$$\begin{aligned}\int_{\mathbb{R}} W(u_\star(x))\, dx &= \int_{-\infty}^{0} \big(W(u_\star(x)) - W(u_\star(-\infty))\big)\, dx + \int_{0}^{+\infty} \big(W(u_\star(x)) - W(u_\star(+\infty))\big)\, dx \\ &\leqslant C \left(\int_{-\infty}^{-R} |u_\star(x) - u_\star(-\infty)|^2\, dx + \int_{R}^{+\infty} |u_\star(x) - u_\star(+\infty)|^2\, dx + 1 \right) \\ &\leqslant C \left(\int_{(-\infty,-R)\cup(R,+\infty)} \frac{dx}{|x|^{4s}} + 1 \right) < +\infty.\end{aligned}$$

obtaining uniform estimates via elliptic theory (actually, via a combination of classical and fractional elliptic theories). For this, an important step is to detect how the constrained minimizers detach from the windows: namely, since the energy by itself cannot guarantee continuity in this fractional exponent range, one needs an auxiliary argument to avoid that minimizers "jump" away when touching the boundary of the constraints. These detachment estimates are obtained from the analysis of the corresponding obstacle problem (the window playing the role of the obstacle at detaching points).

Another technical issue in this fractional exponent range is that approximating heteroclinics may oscillate between equilibria when placing the windows far from each other: to avoid this pathology, we use a second penalization method to charge the L^2-difference with a centered excursion, so to control the "baricenter" of the heteroclinic.

3. Dislocation dynamics

To understand the way in which dislocations move, one can look at the parabolic analogue of equation (1.7), namely consider solutions $v = v(t, x)$ of

$$\partial_t v = -(-\Delta)^s v - W'(v) \quad \text{in } (0, +\infty) \times \mathbb{R}. \tag{3.1}$$

More generally, one can consider the equation

$$\partial_t v = -(-\Delta)^s v - W'(v) + \sigma_\varepsilon,$$

where σ_ε is a suitably small external stress, but to keep the exposition as simple as possible we take here $\sigma_\varepsilon \equiv 0$.

In this section, we recall some results aiming at describing the evolution of the parabolic dislocation function v for suitably large space and time scales. To accomplish this goal, it is appropriate to scale both the space and the time variables according to the homogeneity of the equation, that is considering the function

$$v_\varepsilon(t, x) := v\left(\frac{t}{\varepsilon^{1+2s}}, \frac{x}{\varepsilon}\right). \tag{3.2}$$

Here $\varepsilon > 0$ is a small parameter: sending $\varepsilon \searrow 0$ would thus correspond to considering large time and space scales, in a suitably intertwined manner that detects a coherent pattern.

By scaling (3.1), one obtains the evolution equation for v_ε given by

$$\varepsilon \partial_t v_\varepsilon = -(-\Delta)^s v_\varepsilon - \frac{1}{\varepsilon^{2s}} W'(v_\varepsilon) \quad \text{in } (0, +\infty) \times \mathbb{R}. \tag{3.3}$$

Since the dynamics of the atom dislocations can be quite complicated, it is also convenient to take into account "well-prepared" initial data, obtained by the superposition of transition layers. That is, we consider the heteroclinic solution $u_\star$ constructed in [**PSV13, CS15**] and mentioned in (2.3) (normalized such that $u_\star(0) = \frac{1}{2}$): this will be the transition layer (from 0 to 1) used as a building block for our construction. It is also convenient to consider a "reversed" transition layer (from 1 to 0) given by $x \mapsto u_\star(-x)$. Hence, we take as initial datum for v a superposition of transition layers and reversed transition layers centered[7] at given points $\bar{x}_1 < \cdots < \bar{x}_N$.

[7] More explicitly, the "center" of a transition layer going from 0 to 1 is the point at which it attains value $\frac{1}{2}$. In particular, we have that $u_\star$ is centered at 0 and $u_\star(x - \bar{x})$ is centered at $\bar{x}$.

From the physical point of view, one can imagine that the points $\bar{x}_i$ represent the initial location of the atom dislocations: the initial configuration is thus "locally" sufficiently close to an equilibrium, as described by the layer centered at $\bar{x}_i$ (the other approaching a constant away from their own center); notice that in this initial setting the atoms are located to either the right or the left of their natural rest position, according to the choice of layer or reversed layer.

To formalize this aspect (that is, to distinguish between transition layers and reversed transition layers), we let $\zeta_1, \ldots, \zeta_N \in \{-1, 1\}$. Roughly speaking, each ζ_i denotes the "orientation" of the dislocation located at $\bar{x}_i$, with $\zeta_i = 1$ corresponding to a layer transition and $\zeta_i = -1$ to a reversed one.

With this notation, the initial datum for equation (3.3) is taken of the form

$$\bar{v}_\varepsilon(x) := v_\varepsilon(0, x) := \sum_{i=1}^{N} u_\star\left(\frac{\zeta_i}{\varepsilon}(x - \bar{x}_i)\right) - K, \tag{3.4}$$

where we denote by $K \in \{1, \ldots, N\}$ the number of negative orientations, i.e.

$$K := \#\{i \text{ s.t. } \zeta_i = -1\}.$$

The reason for subtracting the constant K in (3.4) is that in this way if we have N even and $K = N/2$ the function in (3.4) is normalized such that $v_\varepsilon(-\infty) = v_\varepsilon(+\infty) = 0$ (we will come back to this normalizing choice on page 125).

The main result in this setting is that, as $\varepsilon \searrow 0$, the dislocations have the tendency to concentrate at single points of the crystal, where the size of the slip coincides with the natural periodicity of the medium: more explicitly, as $\varepsilon \searrow 0$, we have that v_ε approaches a step function v_0 of the form

$$v_0(t, x) := \sum_{i=1}^{N} H\left(\zeta_i(x - x_i(t))\right) - K, \tag{3.5}$$

where H is the Heaviside function (and we stress that the jump of the Heaviside function is unitary, hence corresponding to the lattice periodicity).

We stress that (3.3) is a "singular perturbation problem" for as much as the small parameter ε directly affects the differential operator (in particular, sticking $\varepsilon = 0$ into (3.3) would be too inaccurate). Similarly, we remark that the function in (3.5) is discontinuous, whence it is not a solution of (3.1).

The entity $x_i(t)$ in (3.5) describes the discontinuities of the "limit dislocation function" as the time t varies, that is, it outlines the dynamics of the dislocation in time.

We have that the motion of these dislocation points is governed by an interior potential. This potential is either repulsive or attractive, depending on the mutual orientations of the dislocations. More precisely, the dislocation dynamics described by $x_i(t)$ correspond to the solutions of the Cauchy problem of a system of ordinary differential equations of the form

$$\begin{cases} \dot{x}_i(t) = \gamma \sum_{\substack{1 \leqslant j \leqslant N \\ j \neq i}} \zeta_i \zeta_j \frac{x_i(t) - x_j(t)}{2s\,|x_i - x_j|^{1+2s}}, \\ x_i(0) = \bar{x}_i, \end{cases} \tag{3.6}$$

where

$$\gamma := \frac{1}{\|u_\star'\|_{L^2(\mathbb{R})}}. \tag{3.7}$$

Interestingly, the system in (3.6) is a gradient flow. We stress that when $\zeta_i\zeta_j = 1$, the potential term is of repulsive type, but when $\zeta_i\zeta_j = -1$ the term is attractive and this may lead to collisions, as we will discuss in Section 4: in particular, one can define $T_c \in (0, +\infty]$ to be the first time when a collision occurs.

Summarizing, by (3.5) and (3.6), at an appropriately large space and time scale, the behavior of the dislocations is governed by an evolving step function (namely the one in (3.5)) and a system of ordinary differential equations describing its jump discontinuities[8] and accounting for the dynamics of the dislocation points (as stated explicitly in (3.6)).

The precise mathematical formulation of this phenomenon goes as follows:

THEOREM 2. *For every $\varepsilon > 0$, there exists a unique viscosity solution of* (3.3) *in* $[0, +\infty) \times \mathbb{R}$, *with initial datum* (3.4).

As $\varepsilon \searrow 0$, this solution exhibits the following asymptotic behavior for all $(t, x) \in [0, T_c) \times \mathbb{R}$, for some $T_c \in (0, +\infty]$:

$$\begin{aligned} &\limsup_{\substack{(t',x')\to(t,x)\\ \varepsilon\searrow 0}} v_\varepsilon(t', x') \leqslant \overline{v_0}(t, x) \\ \text{and} \quad &\liminf_{\substack{(t',x')\to(t,x)\\ \varepsilon\searrow 0}} v_\varepsilon(t', x') \geqslant \underline{v_0}(t, x), \end{aligned} \tag{3.8}$$

where v_0 is as in (3.5),

$$\begin{aligned} \overline{v_0}(t, x) &:= \limsup_{(t',x')\to(t,x)} v_0(t', x') \\ \text{and} \quad \underline{v_0}(t, x) &:= \liminf_{(t',x')\to(t,x)} v_0(t', x'). \end{aligned} \tag{3.9}$$

Moreover, the orbits $x_i(t)$ in the definition of v_0 in (3.5) *are solutions of the Cauchy problem in* (3.6).

We observe that outside the jumps $x_i(t)$, the function v_0 is continuous, hence (3.9) boils down to

$$\overline{v_0}(t, x) = \underline{v_0}(t, x) = v_0(t, x) \quad \text{provided that } x \notin \{x_1(t), \dots, x_N(t)\},$$

and correspondingly (3.8) reduces to

$$\lim_{\substack{(t',x')\to(t,x)\\ \varepsilon\searrow 0}} v_\varepsilon(t', x') = v_0(t, x) \quad \text{provided that } x \notin \{x_1(t), \dots, x_N(t)\}. \tag{3.10}$$

In this sense, the statement in (3.8) is a refinement of the simpler one in (3.10) in which one takes into account the upper and lower semi-continuous envelopes of v_0 to have a result comprising also the discontinuity set of the limit dislocation function.

Theorem 2 was first proved in [**GM12**] when $s = \frac{1}{2}$ (compare with footnote 5), and assuming that $\zeta_1 = \cdots = \zeta_N = 1$ (the case $\zeta_1 = \cdots = \zeta_N = -1$ being analogous).

The case $s \in \left(\frac{1}{2}, 1\right)$ with $\zeta_1 = \cdots = \zeta_N = 1$ has been established in [**DPV15**] and the case $s \in \left(0, \frac{1}{2}\right)$ with $\zeta_1 = \cdots = \zeta_N = 1$ in [**DFV14**].

---

[8]On the one hand, we note that since v_0 is discontinuous, its asymptotic behavior for large times cannot be directly related to the steady state solutions in (1.6). On the other hand, we will reconsider this possibility by taking into account a different scale in Section 5: namely, instead of sending $\varepsilon \searrow 0$ right away and then considering long time asymptotics, in Section 5 we will address the long time asymptotic regime for the case of a small parameter $\varepsilon > 0$; in this scaling, the role of the transition layers will become handy and expedient.

Gradient flows as in (3.6) with $\zeta_1 = \cdots = \zeta_N = 1$ were studied in [**FIM09**].

Theorem 2 in its full generality was then proved in [**PV15a**], where the case of collisions for the Cauchy problem in (3.6) under opposite orientations was also taken into account.

4. Collisions

In light of the discussion in Section 3, a natural question is to analyze the collision time T_c in Theorem 2, that is the first time in which two trajectories of the system of ordinary differential equations in (3.6) meet at the same point, namely

$$\begin{gathered} x_i(t) \neq x_j(t) \text{ for all } i \neq j \text{ when } t \in [0, T_c), \\ x_{i_0}(T_c) = x_{i_0+1}(T_c) \text{ for some } i_0 \in \{1, \dots, N-1\}. \end{gathered}$$

In particular, the system in (3.6) cannot be extended for $t > T_c$. We stress that T_c is finite only in presence of different orientations, since when the orientations ζ_i are all positive, or all negative, then the potential is repulsive and particles do not collide.

The collision time T_c can be explicitly estimated in several concrete cases, such as the one with two dislocations with opposite orientations, that of three dislocations with alternate orientations, that of N dislocations with alternate orientations, etc.

For instance, among the several estimates obtained in [**PV15a**], we recall:

THEOREM 3.

(i). *Let $N = 2$ and $K = 1$. Then,*

$$T_c \leqslant \frac{s(\bar{x}_2 - \bar{x}_1)^{2s+1}}{(2s+1)\gamma},$$

where γ is as in (3.7).

(ii). *Let $N = 3$, $\zeta_1 = \zeta_3 = 1$ and $\zeta_2 = -1$. Let also*

$$\tau := \frac{s\big(\min\{\bar{x}_2 - \bar{x}_1,\, \bar{x}_3 - \bar{x}_2\}\big)^{2s+1}}{(2s+1)\gamma} \qquad \textit{and} \qquad C := \frac{2^{2s+1}}{2^{2s}-1}.$$

Then,

$$T_c \in [\tau, C\tau].$$

(iii). *Let $N \geqslant 2$ and $K \leqslant N-1$, with $\zeta_{i_0}\zeta_{i_0+1} = -1$ for some $i_0 \in \{1, \dots, N-1\}$. Then, there exists $a_0 \in \left(0, \frac{1}{(N-2)^{\frac{1}{2s}}}\right)$ such that the following statement holds true.*

Assume that

$$\bar{x}_{i_0+1} - \bar{x}_{i_0} \leqslant a_0 \min_{\substack{i \in \{1,\dots,N-1\} \\ i \neq i_0}} \bar{x}_{i+1} - \bar{x}_i.$$

Then,

$$x_{i_0+1}(t) - x_{i_0}(t) \leqslant a_0 \min_{\substack{i \in \{1,\dots,N-1\} \\ i \neq i_0}} x_{i+1}(t) - x_i(t)$$

for all $t \in [0, T_c)$, and

$$T_c \leqslant \frac{s(\bar{x}_{i_0+1} - \bar{x}_{i_0})^{2s+1}}{(2s+1)\gamma(1-(N-2)a_0^{2s})}.$$

(iv). *Let* $N \geqslant 2$ *and* $\zeta_i\zeta_{i+1} = -1$ *for all* $i \in \{1, \ldots, N-1\}$.
Then,

$$T_c \leqslant \begin{cases} \frac{(N-1)(\bar{x}_N - \bar{x}_1)^{2s+1}}{(2s+1)\gamma} & \text{if } N \text{ is odd,} \\ \frac{s(\bar{x}_N - \bar{x}_1)^{2s+1}}{(2s+1)\gamma} & \text{if } N \text{ is even.} \end{cases}$$

In particular, Theorem 3(i) deals with the case of two dislocations with opposite orientation and detects the collision time in terms of their mutual initial distance: as a matter of fact, the result in [**PV15a**] is more general than this, since it takes into account also a possible external stress, in which case the result mentioned here holds true provided that the dislocations are initially positioned sufficiently close to each other (otherwise we also show that there are cases in which $T_c = +\infty$).

Theorem 3(ii) considers the case of three alternate dislocations and provides bounds from above and below on the collision time. In this situation, if two dislocations, say the first and the second, are closer than the others, this property remains true for all times till collision occur: namely, if $\bar{x}_2 - \bar{x}_1 < \bar{x}_3 - \bar{x}_2$ then $x_2(t) - x_1(t) < x_3(t) - x_2(t)$ for all $t \in [0, T_c)$.

This property is also useful to characterize "triple collisions", that is the situation in which $x_1(T_c) = x_2(T_c) = x_3(T_c)$. In the setting of Theorem 3(ii), one has that triple collisions occur if and only if the initial distance between the dislocation points is the same and in this case one can characterize explicitly the collision time as the larger possible time obtained in Theorem 3(ii). More precisely, triple collisions occur if and only if

$$\bar{x}_2 - \bar{x}_1 = \bar{x}_3 - \bar{x}_2 \qquad \text{and} \qquad T_c = C\tau.$$

Theorem 3(iii) deals with the general case of N dislocations, assuming that the orientation of the $(i_0 + 1)$th and the one of the i_0th are opposite. In this situation, if the initial distance between these two dislocation points is sufficiently small with respect to the others, then so it remains for all times till collision occurs, and one can also give an explicit bound on the collision time.

Finally, Theorem 3(iv) considers the case of N dislocations with alternate orientations and bounds the collision time in terms of the maximal distance between initial dislocation points.

The estimates in Theorem 3 are essentially sharp. For instance, these estimates relate the collision times to the power $2s + 1$ of the maximal dislocation distance. This feature is optimal, see the explicit solutions exhibited on pages 243–246 in [**PV15a**].

5. Relaxation times and large time asymptotics

Since the dawn of history, human beings have asked some fundamental questions, such as "what happens after death?". Being unable to address this question, we focus on the weaker problem of discussing what happens to the dislocations after collisions.

For this, we formally relate the collision time T_c discussed in Section 4 with the asymptotics presented in Theorem 2. Let us focus for simplicity on the case of two dislocations with opposite orientations, i.e. $N = 2$ and $K = 1$, say with $\zeta_1 = 1$ and $\zeta_2 = -1$, and let $x_c := x_1(T_c) = x_2(T_c)$ denote the collision point. Thus, if $x \neq x_c$ and t is sufficiently close to T_c, we have that $x_1(t) \neq x$ and $x_2(t) \neq x$, whence it follows from (3.5) and (3.10) that, when t is sufficiently close to T_c,

$$\lim_{\varepsilon\searrow 0} v_\varepsilon(t,x) = v_0(t,x) = \sum_{i=1}^{N} H\left(\zeta_i(x - x_i(t))\right) - 1 = H(x - x_1(t)) + H(x_2(t) - x) - 1.$$

In this setting, when t is sufficiently close to T_c, we have that either $x < x_1(t) < x_2(t)$ (giving that $H(x - x_1(t)) = 0$ and $H(x_2(t) - x) = 1$) or $x_1(t) < x_2(t) < x$ (giving that $H(x - x_1(t)) = 1$ and $H(x_2(t) - x) = 0$). From these observations we conclude that

$$\lim_{\varepsilon\searrow 0} v_\varepsilon(t,x) = 0$$

as long as t is sufficiently close to T_c, whence

$$\lim_{t\nearrow T_c}\lim_{\varepsilon\searrow 0} v_\varepsilon(t,x) = 0. \tag{5.1}$$

However, in [**PV15a**] it is also proved that

$$\limsup_{\substack{t\nearrow T_c\\ \varepsilon\searrow 0}} v_\varepsilon(t,x_c) \geqslant 1. \tag{5.2}$$

This says that while, by (5.1), the two dislocations with opposite orientation average after the collision, the dislocation function at the collision point and at the collision time "keeps a bit of a memory" of the collision event, as given by (5.2).

In view also of this phenomenon, it is interesting to detect the "relaxation time" of the dislocation after collision. This question was investigated in [**PV16**] in which it is established that, after a small transition time subsequent to the collision, the dislocation function relaxes exponentially fast to a steady state. In this sense, the memory effect at the collision time detected in (5.2) fades very fast. Moreover, the system exhibits two separate scales of decay behaviors, namely an exponentially fast decay in time and a polynomial decay in the space variable. It is interesting that these two scales are somewhat asymptotically kept separate: roughly speaking, for large times the main contribution comes from the time derivative of the dislocation function balancing the potential (which produces an ordinary differential equation in time leading to exponential decay), while in the space variable the leading term comes from the balance of the fractional Laplacian of the dislocation function with the potential (which, as in (2.4), provides a slow decay of polynomial type).

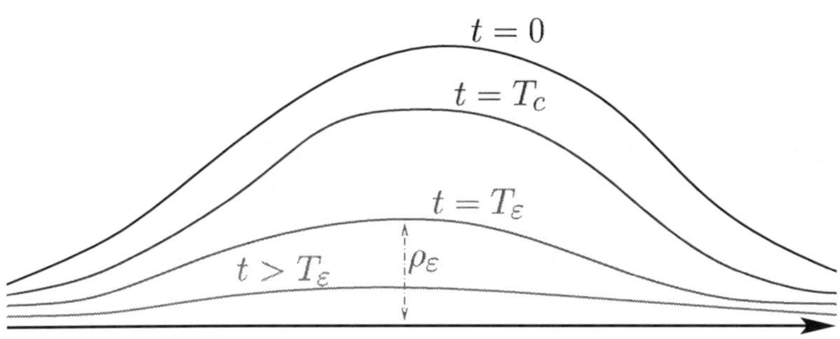

FIGURE 18. The two dislocations situation described in Theorem 4.

The precise result for the two dislocations collision go as follows (see Figure 18 for a sketch of this situation):

THEOREM 4. *Let $s \in (0,1)$. Let $N = 2$ and $K = 1$. Then there exist $\varepsilon_0 > 0$ and $c > 0$ such that for all $\varepsilon \in (0, \varepsilon_0)$ we have that, for all $t \geqslant T_\varepsilon$,*

$$|v_\varepsilon(t,x)| \leqslant \rho_\varepsilon \exp\left(-\frac{c}{\varepsilon^{2s+1}}(t - T_\varepsilon)\right),$$

where $\rho_\varepsilon \searrow 0$ and $T_\varepsilon \searrow T_c$ as $\varepsilon \searrow 0$.

The case of three alternate dislocations is similar, albeit different, since the steady state associated with this case is a heteroclinic and not the zero function. That is, while in the case of two dislocations the collision leads to an "annihilation to the trivial equilibrium", in the case of three dislocations we have that two dislocations "annihilate each other", but the third one "survives" and dictates the asymptotic behavior (see Figure 19 for a sketch of this situation). The precise result goes as follows and provides a trapping from above and below for the solution v_ε in terms of the heteroclinic connection $u_\star$:

THEOREM 5. *Let $s \in (0,1)$. Let $N = 3$, $\zeta_1 = \zeta_3 = 1$ and $\zeta_2 = -1$. Then there exist $\varepsilon_0 > 0$ and $c > 0$ such that for all $\varepsilon \in (0, \varepsilon_0)$ we have that, for all $t \geqslant T_\varepsilon$,*

$$v_\varepsilon(t,x) \leqslant u_\star\left(\frac{1}{\varepsilon}\left(x - y_\varepsilon + \kappa_\varepsilon\rho_\varepsilon\left(1 - \exp\left(-\frac{c}{\varepsilon^{2s+1}}(t - T_\varepsilon)\right)\right)\right)\right) + \rho_\varepsilon \exp\left(-\frac{c}{\varepsilon^{2s+1}}(t - T_\varepsilon)\right)$$

and

$$v_\varepsilon(t,x) \geqslant u_\star\left(\frac{1}{\varepsilon}\left(x - z_\varepsilon - \kappa_\varepsilon\rho_\varepsilon\left(1 - \exp\left(-\frac{c}{\varepsilon^{2s+1}}(t - T_\varepsilon)\right)\right)\right)\right) - \rho_\varepsilon \exp\left(-\frac{c}{\varepsilon^{2s+1}}(t - T_\varepsilon)\right),$$

where $\rho_\varepsilon \searrow 0$, $\kappa_\varepsilon \searrow 0$, $T_\varepsilon \searrow T_c$ and y_ε, $z_\varepsilon \to x_\star$, for some $x_\star \in \mathbb{R}$, as $\varepsilon \searrow 0$.

In particular,

$$\limsup_{t\to+\infty} v_\varepsilon(t,x) \leqslant u_\star\left(\frac{1}{\varepsilon}(x - y_\varepsilon + \kappa_\varepsilon\rho_\varepsilon)\right)$$

and

$$\liminf_{t\to+\infty} v_\varepsilon(t,x) \geqslant u_\star\left(\frac{1}{\varepsilon}(x - z_\varepsilon - \kappa_\varepsilon\rho_\varepsilon)\right).$$

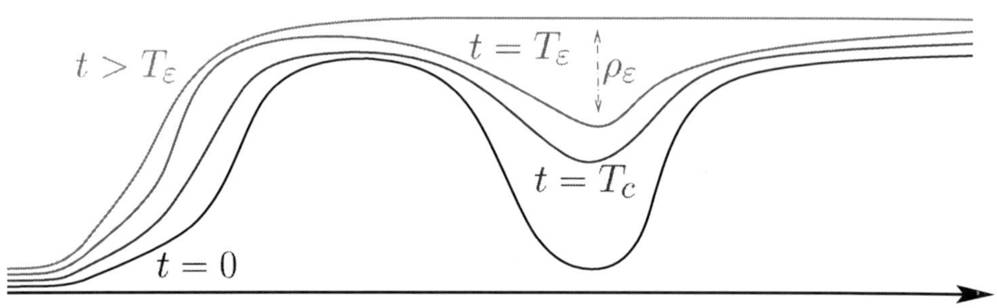

FIGURE 19. The three alternate dislocations situation described in Theorem 5.

The case of N dislocations has been discussed in [**PV17**]. The general feature, besides technicalities, is that the parabolic character of the equation entails a "smoothing effect" on the dislocation function slightly after collision, which leads, for large time, to a precise asymptotics to either a constant function or a single heteroclinic, depending on the algebraic properties of the orientations of the initial datum.

To address the problem in its full complexity, it is convenient to distinguish the case in which the dislocations are "balanced", i.e. N is even and $K = N/2$, corresponding to the situation in which the number of positively oriented dislocation is equal to that of the negatively oriented ones, and the case in which the dislocations are "unbalanced", namely the number of positive ones exceeds that of negative ones (or vice versa). A heuristic picture depicting the balanced case $N = 4$ and $K = 2$, with $\zeta_1 = \zeta_3 = 1$ and $\zeta_2 = \zeta_4 = -1$ is sketched[9] in Figure 20.

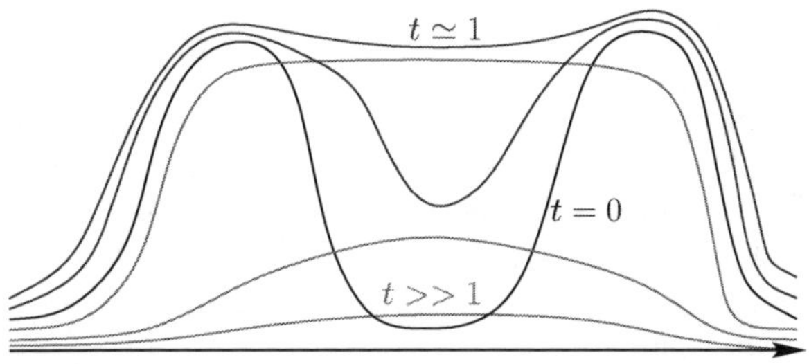

FIGURE 20. Time asymptotic for the balanced case.

In the balanced regime, given the normalization in (3.4), we know that, the initial dislocation function $\bar{v}_\varepsilon$ goes to zero both as $t \to -\infty$ and as $t \to +\infty$ and it turns out that this asymptotic behavior in space also influences the asymptotic behavior in time. More precisely, we have that after a transient time in which collisions occur, the dislocation function v_ε relaxes to zero exponentially fast.

The unbalanced case requires a more specific analysis. In this case, we can suppose that $K < N - K$ (the case $K > N - K$ being similar) and define $\ell := N-2K > 0$. We observe that ℓ counts the difference between the positively oriented initial transitions (which are $N-K$) and the negatively oriented ones (which are K). In this case, thanks to the normalization in (3.4) the initial datum $\bar{v}_\varepsilon$ is asymptotic to zero as $x \to -\infty$ and to ℓ as $x \to +\infty$.

Given our previous discussion about the balanced case, one can expect that the long time behavior of the dislocation function is influenced by these initial conditions and one might believe that, as $t \to +\infty$ the system will try to average between the two values 0 and ℓ. However, this has to be taken with a pinch of salt, since the constant $\frac{\ell}{2}$ (that is the above mentioned average) is not a minimum of the potential W, unless ℓ is even, and therefore we cannot expect that v_ε converges to $\frac{\ell}{2}$ as $t \to +\infty$, unless ℓ is even.

The case ℓ odd needs therefore an additional analysis. On the one hand, in principle, the constant $\frac{\ell}{2}$ could be a stationary point for W, hence it could provide a solution of the corresponding stationary equation in (1.7). On the other hand, even if this happened, the corresponding solution would be unstable from the variational point of view, hence not suited to attract the dynamics of the parabolic flow.

[9]Coming back to footnote 8, it is suggestive to think about Figure 20 as a backbone structure for different collisions happening in the background, either simultaneously (in case of a fully symmetric picture) or at different times. However, we also stress that the situation taken into account in Figure 20 is somewhat different from that presented in Section 4, since here the parameter ε is small, but positive, and we consider arbitrarily long times, while in Section 4 the order in which we take limits is the other way around, namely in that case one first considers the limit as $\varepsilon \searrow 0$ and then looks at the long time asymptotics. Since the problem is singular, the order in which these limits is performed does play a role in the rigorous descriptions of the results.

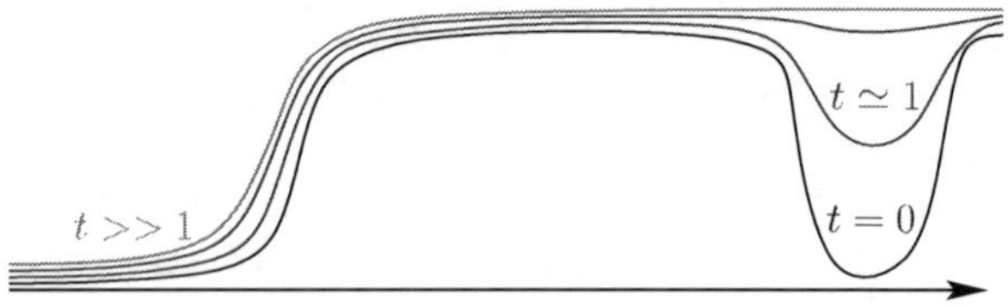

FIGURE 21. Time asymptotic for the unbalanced case (with $N = 3$ and $K = 1$).

As a result, when ℓ is odd the average constant $\frac{\ell}{2}$ is not the asymptotic limit of v_ε as $t \to +\infty$, but it does serve as average asymptotics between the values $\frac{\ell-1}{2}$ and $\frac{\ell+1}{2}$, which are now integer and correspond to stable solutions of (1.7) induced by the minima of the potential W. What happens when ℓ is odd is thus that, as $t \to +\infty$, the dislocation function v_ε converges to a transition layer from $\frac{\ell-1}{2}$ to $\frac{\ell+1}{2}$.

For instance, a sketch of the situation in which $N = 3$ and $K = 1$, with $\zeta_1 = \zeta_3 = 1$ and $\zeta_2 = -1$ is given in Figure 21. In this case, we have that $\ell = 3-2 = 1$ hence ℓ is odd and the limit configuration in time is a transition from $\frac{\ell-1}{2} = 0$ to $\frac{\ell+1}{2} = 1$.

Instead, a sketch of the situation in which $N = 4$ and $K = 1$, with $\zeta_1 = \zeta_2 = \zeta_3 = 1$ and $\zeta_4 = -1$ is given in Figure 21. In this case, we have that $\ell = 4-2 = 2$, which is even and the asymptotic state corresponds to the constant $\frac{\ell}{2} = 1$. This asymptotics is indeed reached, roughly speaking, after the collision (and consequent relaxation) of the third and forth dislocation, followed by the repulsive interaction between the first and second dislocation (this repulsion is responsible for the drift towards $-\infty$ and $+\infty$ of these two transition layers).

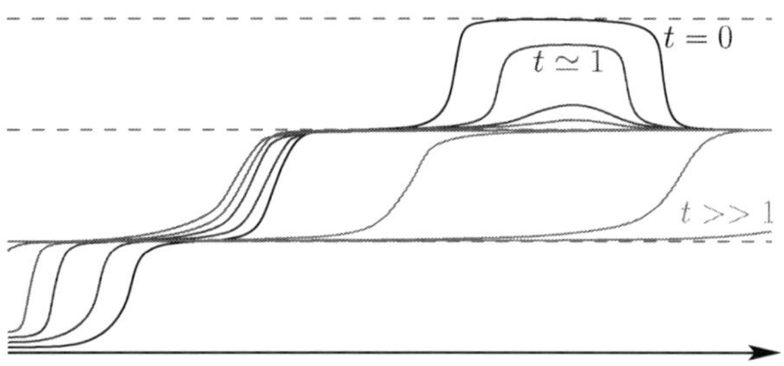

FIGURE 22. Time asymptotic for the unbalanced case (with $N = 4$ and $K = 1$).

See [**PV17**] for the corresponding formal statements. We stress that the pictures of this paper are freehand drawn, not the outcome of a careful numerical simulation, hence their intent is only explanatory, they are at best qualitative and no attempt has been made to reproduce quantitatively in the pictures the different scales involved in the asymptotics.

It is also interesting to compare Figure 22 with the multibump solution in Figure 13. In spite of their apparent pictorial similarity, we stress that an important conceptual difference between the two settings is given by the fact that here we are dealing with solutions of an evolution equation, while Figure 13 represents a steady state. Also, the multibump solution in Figure 13 is produced thanks to a generic perturbation of the potential (recall (2.2)): this perturbation plays a decisive role for the formation of multibumps, since the unperturbed case does not possess arbitrarily oscillatory solutions. The evolutionary and unperturbed case dealt with in this section also does not admit oscillatory solutions in the long time asymptotics, since, as established in [**PV17**], the large time profile in this situation is either a constant or a single heteroclinic connecting two integer values. This is the main reason for which the oscillations appearing in Figure 22 for small time end up "annihilating each other" as time flows (with a mechanisms that heuristically reminds the collision framework discussed in Section 4, but with the caveat that here $\varepsilon > 0$).

6. Orowan's Law

In this section we will focus on the physical case $s = \frac{1}{2}$ and we will deduce the so-called Orowan's Law, stating that the plastic strain velocity is proportional to the product of the density of dislocations with the effective stress.

In Section 3 we have seen that the dynamics of N parallel straight dislocation lines, all lying in the same slip plane, and all positive oriented ($\xi_i = 1$ for $i = 1, \dots, N$) can be described by the ODE's system

$$\begin{cases} \dot{x}_i(t) = \gamma \sum_{\substack{1\leqslant j\leqslant N\\ j\neq i}} \frac{1}{x_i - x_j}, \\ x_i(0) = \bar{x}_i, \end{cases} \tag{6.1}$$

recall (3.6) and (3.7).

Here $T_c = +\infty$ as points with same orientation repeal each other, thus never collide. In this setting, we are interested in the case when the number N of dislocation points goes to infinity. We will present below some results proven in [**PS21**] and aiming at identifying at large scale an evolution model for the dynamics of a density of dislocations. For this, let us introduce a new parameter $\delta = \delta_\varepsilon$ such that $\delta \to 0$ as $\varepsilon \to 0$ and let us consider the solution v_ε of (3.3) with initial condition (3.4) (with $s = \frac{1}{2}$ and $\xi_i = 1$), where now

$$N = N_\delta \sim \frac{1}{\delta} \to +\infty \quad \text{as } \varepsilon \to 0.$$

We next consider the following rescaling

$$u_\varepsilon(t,x) := \delta v_\varepsilon\left(\frac{t}{\delta}, \frac{x}{\delta}\right).$$

Then u_ε is solution to

$$\varepsilon \partial_t u_\varepsilon = -(-\Delta)^{\frac{1}{2}} u_\varepsilon - \frac{1}{\varepsilon} W'\left(\frac{u_\varepsilon}{\delta}\right) \quad \text{in } (0,+\infty)\times\mathbb{R} \tag{6.2}$$

with initial condition

$$u_\varepsilon(0,x) = \sum_{i=1}^{N_\delta} \delta u_\star\left(\frac{x - \bar{y}_i}{\varepsilon\delta}\right) \quad \text{on } \mathbb{R}, \tag{6.3}$$

where, for $i = 1, \dots, N_\delta$,

$$\bar{y}_i := \delta \bar{x}_i$$

and $u_\star$ is the transition layer introduced in (2.3).

Notice that with this rescaling $u_\varepsilon(0,x)$ (and thus, by comparison principle, $u_\varepsilon(t,x)$) is bounded uniformly in ε as $|u_\varepsilon(0,x)| \leqslant N_\delta \delta \sim 1$.

The initial condition (6.3) can be replaced, in further generality, by any non-decreasing $C^{1,1}$ function u_0. On the other hand, any function $v \in C^{1,1}(\mathbb{R})$ which is non-decreasing and such that $v(-\infty) = 0$, can be approximated by a function of the form (6.3). Indeed, given such a v, for $0 < \delta < 1$, define the δi level set points y_i as follows

$$y_i := \inf\{x \in \mathbb{R} \text{ s.t. } v(x) = \delta i\} \quad \text{for all } i = 1, \dots, N_\delta, \tag{6.4}$$

where

$$N_\delta := \left\lfloor \frac{v(+\infty) - \delta}{\delta} \right\rfloor.$$

Since the function v is non-decreasing, we have that $y_i < y_{i+1}$. Then, v can be approximated in $L^\infty(\mathbb{R})$ as stated in the following result.

PROPOSITION 6. *Let $v \in C^{1,1}(\mathbb{R})$ be non-decreasing, non-constant and such that $v(-\infty) = 0$. Let y_i be defined as in (6.4). Then, for all $x \in \mathbb{R}$,*

$$\left| \sum_{i=1}^{N_\delta} \delta u_\star \left(\frac{x - y_i}{\varepsilon \delta} \right) - v(x) \right| \leqslant o_{\varepsilon,\delta}(1),$$

where $o_{\varepsilon,\delta}$ is independent of x and $o_{\varepsilon,\delta}(1) \to 0$ as $\varepsilon, \delta \to 0$.

Moreover, for v and y_i as above, the following approximation formula holds true: for any $i = 1, \dots, N_\delta$,

$$(-\Delta)^{\frac{1}{2}} v(y_i) = \sum_{j \neq i} \frac{\delta}{y_i - y_j} + o_\delta(1), \tag{6.5}$$

where $o_\delta(1) \to 0$ as $\delta \to 0$, uniformly on $\mathbb{R}$. Below is the main result in this setting.

THEOREM 7. *Assume that $u_0 \in C^{1,1}(\mathbb{R})$ and u_0 is non-decreasing. Let u_ε be the viscosity solution of (6.2), with initial condition $u_\varepsilon(0,x) = u_0(x)$, for all $x \in \mathbb{R}$. Then, as $\varepsilon \to 0$, u_ε converges locally uniformly in $(0,+\infty) \times \mathbb{R}$ to the unique non-decreasing in x viscosity solution $\bar{u}$ of*

$$\begin{cases} \partial_t u = -\gamma \partial_x u (-\Delta)^{\frac{1}{2}} u & \text{in } (0,+\infty) \times \mathbb{R}, \\ u(0,\cdot) = u_0 & \text{on } \mathbb{R}, \end{cases} \tag{6.6}$$

where γ is as in (3.7)

The limit problem (6.6) represents the plastic flow rule for the macroscopic crystal plasticity with density of dislocations, where the solution $\bar{u}$ can be interpreted as the plastic strain, $(-\Delta)^{\frac{1}{2}} \bar{u}$ is the internal stress created by the density of dislocations contained in a slip plane and identified by $\partial_x \bar{u} \geqslant 0$. Theorem 7 says that, in this regime, the plastic strain velocity is proportional to the dislocation density times the effective stress. This physical law is known as Orowan's equation, see e.g. page 3739 in [**RSJKEW03**].

Equation

$$\partial_t \bar{u} = -\gamma \partial_x \bar{u} (-\Delta)^{\frac{1}{2}} \bar{u} \tag{6.7}$$

is an integrated form of a model studied by Head [**Hea72**] for the self-dynamics of the dislocation density $\partial_x \bar{u}$. Indeed, denoting by $f := \partial_x \bar{u}$, differentiating (6.7), we see that, at least formally, f solves

$$\partial_t f = c_0 \partial_x (f \mathscr{H}[f]) \tag{6.8}$$

where $\mathscr{H}$ is the Hilbert transform defined by

$$\mathscr{H}[f](x) := PV \int_{\mathbb{R}} \frac{f(y)}{y-x} dy.$$

Here we used that if $\bar{u} \in C^{1,\alpha}(\mathbb{R})$ and $f = \partial_x \bar{u} \in L^p(\mathbb{R})$ with $1 < p < +\infty$, then an integration by parts yields

$$-(-\Delta)^{\frac{1}{2}} \bar{u} = \mathscr{H}[f].$$

The conservation of mass satisfied by the positive integrable solutions of (6.8) reflects the fact that if $f = \partial_x \bar{u}$ is the density of dislocations, no dislocations are created or annihilated.

Equation (6.8), called by Head the "equation of motion of the dislocation continuum", was obtained by assuming that the speed of each dislocation is proportional to the effective stress at the dislocation.

This physical property is encoded in the limit problem (6.6). To see this, we first observe that the dislocation points can be approximated by the level set points of the limit function $\bar{u}$. More precisely, define $y_i(t)$ as the δi-level set of $\bar{u}(t,\cdot)$ as in (6.4). Assuming that $\bar{u}$ is smooth and $\partial_x \bar{u}(t, y_i(t)) > 0$, we see by differentiating in time the equation $\bar{u}(t, y_i(t)) = \delta i$ and using (6.5) and (6.6), that

$$\dot{y}_i = -\frac{\partial_t \bar{u}(t, y_i(t))}{\partial_x \bar{u}(t, y_i(t))} = \gamma(-\Delta)^{\frac{1}{2}} \bar{u}(t, y_i(t)) \simeq \gamma \sum_{j \neq i} \frac{\delta}{y_i - y_j}. \tag{6.9}$$

Scaling back to $x_i(t) := y_i(\delta t)/\delta$, we get

$$\dot{x}_i \simeq \gamma \sum_{j \neq i} \frac{1}{x_i - x_j},$$

which is, up to small errors, (6.1). Moreover, from (6.9), we see that the mean velocity of each dislocation is proportional to the effective stress $(-\Delta)^{\frac{1}{2}} \bar{u}$ at the dislocation, as assumed by Head.

One can also take into account the case where dislocation points can be either positive or negative oriented, by removing the assumption that the initial datum u_0 is non-decreasing. In this situation Theorem 7 is generalized in a forthcoming paper [**PS**] as follows.

THEOREM 8. *Assume that $u_0 \in C^{1,1}(\mathbb{R})$ and that the limits $u(+\infty)$ and $u(-\infty)$ exist. Let u_ε be the viscosity solution of (6.2), with initial condition $u_\varepsilon(0,x) = u_0(x)$, for all $x \in \mathbb{R}$. Then, as $\varepsilon \to 0$, u_ε converges locally uniformly in $(0,+\infty) \times \mathbb{R}$ to the viscosity solution $\bar{u}$ of*

$$\begin{cases} \partial_t u = -\gamma |\partial_x u| (-\Delta)^{\frac{1}{2}} u & \text{in } (0,+\infty) \times \mathbb{R}, \\ u(0,\cdot) = u_0 & \text{on } \mathbb{R}, \end{cases} \tag{6.10}$$

where γ is as in (3.7).

Differentiating equation $\partial_t \bar{u} = -\gamma |\partial_x \bar{u}| (-\Delta)^{\frac{1}{2}} \bar{u}$ yields the two following equations for the positive and negative part of $f = \partial_x \bar{u}$,

$$\partial_t f^+ = \partial_x (f^+ \mathscr{H}(f^+ - f^-)), \quad \partial_t f^- = -\partial_x (f^- \mathscr{H}(f^+ - f^-))$$

which are the 1-D version of the 2-D Groma-Balogh equations [**GB99**], a macroscopic model describing the evolution of the density of positive and negative oriented parallel straight dislocation lines. In this model in which collisions may occur, the mass of the dislocation density $|\partial_x \bar{u}| = f^+ + f^-$ is not in general preserved, while it is preserved the mass of $f^+ - f^-$, more precisely: for all $t > 0$, $\bar{u}(t, +\infty) = u_0(+\infty)$ and $\bar{u}(t, -\infty) = u_0(-\infty)$, as shown in [**PS**].

7. Homogenization

In this section, we consider equation (6.2) in any dimensions $n \geqslant 1$ and we deal with the corresponding homogenization problem.

To this end, we recall that the parameter ε in (6.2) describes the ratio between the microscopic and the mesoscopic scale, where the discrete dynamics of dislocations is given by the ODE system (3.6). Thus, recalling the initial condition (3.4), the parameter ε can also be interpreted as the distance between dislocation points at microscopic scale.

In this section, we freeze ε, namely we keep this distance fixed but, as in Section 6, we let the number of dislocation points go to infinity. Hence, without loss of generality, we can assume $\varepsilon = 1$. Thus we deal with the problem

$$\begin{cases} \partial_t u_\delta = -(-\Delta)^{\frac{1}{2}} u_\delta - W'\left(\frac{u_\delta}{\delta}\right) & \text{in } (0, +\infty) \times \mathbb{R}^n, \\ u_\delta(0, x) = u_0(x) & \text{on } \mathbb{R}^n. \end{cases} \tag{7.1}$$

The case $n > 1$, in particular $n = 2$, allows us to consider dislocation lines, all lying in the same slip plane, which can be curved and thus for which the reduction of dimension argument described in the introduction cannot be applied.

One could also eventually add on the right hand-side of the PDE in (7.1) a term of the form $\sigma(t/\delta, x/\delta)$, where σ is periodic in both t and x, representing a stress created by the obstacles in the crystal or/and an applied exterior stress. For simplicity of presentation, as before we will take $\sigma \equiv 0$.

The behavior of the solution u_δ of (7.1) has been investigated in [**MP12b**]. From a mathematical viewpoint, problem (7.1) is a homogenization problem, and as usual in periodic homogenization (recall that the potential W is periodic), the limit equation is determined by a cell problem. In this case, such a problem is, for any $p \in \mathbb{R}^n$ and $L \in \mathbb{R}$, the following

$$\begin{cases} \partial_\tau w = -(-\Delta)^{\frac{1}{2}} w + L - W'(w + p \cdot y) & \text{in } (0, +\infty) \times \mathbb{R}^n, \\ w(0, y) = 0 & \text{on } \mathbb{R}^n, \end{cases} \tag{7.2}$$

and one looks for some $\lambda \in \mathbb{R}$ for which $w - \lambda\tau$ is bounded. The precise result goes as follows.

THEOREM 9. *For $L \in \mathbb{R}$ and $p \in \mathbb{R}^n$, there exists a unique viscosity solution $w \in C((0, +\infty) \times \mathbb{R}^n)$ of (7.2) and there exists a unique $\lambda \in \mathbb{R}$ such that w satisfies*

$$\frac{w(\tau, y)}{\tau} \text{ converges towards } \lambda \text{ as } \tau \to +\infty, \text{ locally uniformly in } y.$$

The real number λ is denoted by $\overline{H}(p,L)$. The function $\overline{H}(p,L)$ is continuous on $\mathbb{R}^n \times \mathbb{R}$ and non-decreasing in L.

The function $\overline{H}(p,L)$ defined through the cell-problem (7.2) is called effective Hamiltonian, and the homogenized limit problem is given by

$$\begin{cases} \partial_t u = \overline{H}(\nabla_x u, -(-\Delta)^{\frac{1}{2}} u) & \text{in} \quad (0,+\infty) \times \mathbb{R}^n, \\ u(0,x) = u_0(x) & \text{on} \quad \mathbb{R}^n, \end{cases} \tag{7.3}$$

as stated below.

THEOREM 10. *The solution u_δ of* (7.1) *converges towards the unique viscosity solution u^0 of* (7.3) *locally uniformly in (t,x), as $\delta \searrow 0$, where $\overline{H}$ is defined in Theorem* 9.

As for the limit problems (6.6) and (6.10), the function u^0 can be interpreted as a macroscopic plastic strain satisfying the macroscopic plastic flow rule (7.3). Moreover $(-\Delta)^{\frac{1}{2}} u^0$ is the stress created by the macroscopic density of dislocations, ∇u^0 is the vectorial dislocation density and $|\nabla u^0|$ is the scalar dislocation density.

The behavior of $\overline{H}(p,L)$ for small stress L and small density $|p|$, in dimension $n=1$, has been investigated in [**MP12a**]. In this regime is it shown that

$$\overline{H}(p,L) \simeq \gamma |p| L, \tag{7.4}$$

thus recovering again the Orowan's Law discussed in Section 6. More precisely:

THEOREM 11. *Let p_0, $L_0 \in \mathbb{R}$. Then the function $\overline{H}$ defined in Theorem* 9 *satisfies*

$$\frac{\overline{H}(\varepsilon p_0, \varepsilon L_0)}{\varepsilon^2} \to \gamma |p_0| L_0 \quad \text{as } \varepsilon \to 0^+,$$

where γ is as in (3.7).

When the exterior force σ does not vanish identically but has zero mean value, it is expected, but not proven, a threshold phenomenon as in [**IMR08**], that is

$$\overline{H}(p,L) = 0 \quad \text{if} \quad |L| \quad \text{is small enough.}$$

This means more generally that this homogenization procedure describes correctly the mechanical behaviour of the stress at large scales, but keeps the memory of the microstructure in the plastic law with possible threshold effects.

The results contained in Theorems 10 and 11 have been generalized in [**PV15b**] to the case where $(-\Delta)^{\frac{1}{2}}$ is replaced by $(-\Delta)^s$ with $s \in (0,1)$. The scaling of the system and the results obtained are different according to the fractional parameter $s \in (0,1)$. Namely, when $s > 1/2$ the effective Hamiltonian "localizes" and it only depends on a first order differential operator. Instead, when $s < 1/2$, the non-local features are predominant and the effective Hamiltonian involves the fractional operator of order s. That is, roughly speaking, for any $s \in (0,1)$, the effective Hamiltonian is an operator of order $\min\{2s, 1\}$, which reveals the stronger non-local effects present in the case $s < 1/2$.

The homogenization problems become

$$\begin{cases} \partial_t u_\delta = -\delta^{2s-1}(-\Delta)^s u_\delta - W'\left(\frac{u_\delta}{\delta}\right) & \text{in} \quad (0,+\infty) \times \mathbb{R}^n, \\ u_\delta(0,x) = u_0(x) & \text{on} \quad \mathbb{R}^n, \end{cases} \tag{7.5}$$

for $s > \frac{1}{2}$, and

$$\begin{cases} \partial_t u_\delta = -(-\Delta)^s u_\delta - W'\left(\frac{u_\delta}{\delta^{2s}}\right) + & \text{in} \quad (0,+\infty)\times\mathbb{R}^n, \\ u_\delta(0,x) = u_0(x) & \text{on} \quad \mathbb{R}^n, \end{cases} \tag{7.6}$$

for $s < \frac{1}{2}$.

Notice that the two different scalings formally coincide when $s = \frac{1}{2}$. The solution u_δ of (7.5) converges as $\delta \to 0$ to the solution u^0 of the homogenized problem

$$\begin{cases} \partial_t u = \overline{H}_1(\nabla_x u) & \text{in} \quad (0,+\infty)\times\mathbb{R}^n, \\ u(0,x) = u_0(x) & \text{on} \quad \mathbb{R}^n, \end{cases} \tag{7.7}$$

with an effective Hamiltonian $\overline{H}_1$ which does not depend on the fractional Laplacian anymore, while the solution u_δ of (7.6) converges as $\delta \searrow 0$ to u^0 solution of the following problem

$$\begin{cases} \partial_t u = \overline{H}_2(-(-\Delta)^s u) & \text{in} \quad (0,+\infty)\times\mathbb{R}^n, \\ u(0,x) = u_0(x) & \text{on} \quad \mathbb{R}^n, \end{cases} \tag{7.8}$$

with an effective Hamiltonian $\overline{H}_2$ not depending on the gradient.

The functions $\overline{H}_1$ and $\overline{H}_2$ are determined by the following cell problem: for $p \in \mathbb{R}^n$ and $L \in \mathbb{R}$,

$$\begin{cases} \partial_\tau w = -(-\Delta)^s w + L - W'(w + p\cdot y) & \text{in} \quad (0,+\infty)\times\mathbb{R}^n, \\ w(0,y) = 0 & \text{on} \quad \mathbb{R}^n, \end{cases} \tag{7.9}$$

that is $\overline{H}_1$ and $\overline{H}_2$ are determined by the unique λ for which $w - \lambda\tau$ is bounded (according to the cases $s > \frac{1}{2}$ and $s < \frac{1}{2}$, respectively). More precisely, as before we have:

THEOREM 12. *For $L \in \mathbb{R}$ and $p \in \mathbb{R}^n$, there exists a unique viscosity solution $w \in C((0,+\infty)\times\mathbb{R}^n)$ of (7.9) and there exists a unique $\lambda \in \mathbb{R}$ such that w satisfies:*

$$\frac{w(\tau,y)}{\tau} \text{ converges towards } \lambda \text{ as } \tau \to +\infty, \text{ locally uniformly in } y.$$

The real number λ is denoted by $\overline{H}(p,L)$. The function $\overline{H}(p,L)$ is continuous on $\mathbb{R}^n \times \mathbb{R}$ and non-decreasing in L.

Once the cell problem is solved, the convergence results go as follows:

THEOREM 13. *The solution u_δ of (7.5) converges towards the solution u^0 of (7.7) locally uniformly in (t,x) as $\delta \searrow 0$, where*

$$\overline{H}_1(p) := \overline{H}(p,0)$$

and $\overline{H}(p,L)$ is defined in Theorem 12.

THEOREM 14. *The solution u_δ of (7.6) converges towards the solution u^0 of (7.8) locally uniformly in (t,x) as $\delta \searrow 0$, where*

$$\overline{H}_2(L) := \overline{H}(0,L)$$

and $\overline{H}(p,L)$ is defined in Theorem 12.

The following extension of the Orowan's Law is also proven in [**PV15b**].

THEOREM 15. *Let* p_0, $L_0 \in \mathbb{R}$ *with* $p_0 \neq 0$. *Then the function* $\overline{H}$ *defined in Theorem* 12 *satisfies*

$$\frac{\overline{H}(\delta p_0, \varepsilon^{2s} L_0)}{\varepsilon^{1+2s}} \to \gamma |p_0| L_0 \quad as\ \varepsilon \to 0^+,$$

where γ *is as in* (3.7).

Acknowledgments

The first and third authors are members of AustMS. The first author is supported by the Australian Research Council DECRA DE180100957 "PDEs, free boundaries and applications". The third author is supported by the Australian Laureate Fellowship FL190100081 "Minimal surfaces, free boundaries and partial differential equations".

References

[AV19] Nicola Abatangelo and Enrico Valdinoci, *Getting acquainted with the fractional Laplacian*, Contemporary research in elliptic PDEs and related topics, Springer INdAM Ser., vol. 33, Springer, Cham, 2019, pp. 1–105. MR3967804

[Ame52] S. Amelinckx, *Spiral growth patterns on apatite crystals*, Nature **169** (1952), no. 4307, 841–842, DOI 10.1038/169841a0.

[Bol56] W. Bollmann, *Interference effects in the electron microscopy of thin crystal foils*, Phys. Rev. **103** (1956), 1588–1589, DOI 10.1103/PhysRev.103.1588.

[BV16] Claudia Bucur and Enrico Valdinoci, *Nonlocal diffusion and applications*, Lecture Notes of the Unione Matematica Italiana, vol. 20, Springer, [Cham]; Unione Matematica Italiana, Bologna, 2016, DOI 10.1007/978-3-319-28739-3. MR3469920

[Bur40] J. M. Burgers, *Geometrical considerations concerning the structural irregularities to be assumed in a crystal*, Proc. Phys. Soc. **52** (1940), no. 1, 23–33, DOI 10.1088/0959-5309/52/1/304.

[CS15] Xavier Cabré and Yannick Sire, *Nonlinear equations for fractional Laplacians II: Existence, uniqueness, and qualitative properties of solutions*, Trans. Amer. Math. Soc. **367** (2015), no. 2, 911–941, DOI 10.1090/S0002-9947-2014-05906-0. MR3280032

[CZR91] Vittorio Coti Zelati and Paul H. Rabinowitz, *Homoclinic orbits for second order Hamiltonian systems possessing superquadratic potentials*, J. Amer. Math. Soc. **4** (1991), no. 4, 693–727, DOI 10.2307/2939286. MR1119200

[CDdP20] Matteo Cozzi, Juan Dávila, and Manuel del Pino, *Long-time asymptotics for evolutionary crystal dislocation models*, Adv. Math. **371** (2020), 107242, 109, DOI 10.1016/j.aim.2020.107242. MR4108219

[DFV18] Serena Dipierro, Alberto Farina, and Enrico Valdinoci, *Density estimates for degenerate double-well potentials*, SIAM J. Math. Anal. **50** (2018), no. 6, 6333–6347, DOI 10.1137/17M114933X. MR3890785

[DFV14] Serena Dipierro, Alessio Figalli, and Enrico Valdinoci, *Strongly nonlocal dislocation dynamics in crystals*, Comm. Partial Differential Equations **39** (2014), no. 12, 2351–2387, DOI 10.1080/03605302.2014.914536. MR3259559

[DGV22] Serena Dipierro, Giovanni Giacomin, and Enrico Valdinoci, *Efficiency functionals for the Lévy flight foraging hypothesis*, J. Math Biol. **85** (2022), no. 4, DOI 10.1007/s00285-022-01808-1. MR4484924

[DPV15] Serena Dipierro, Giampiero Palatucci, and Enrico Valdinoci, *Dislocation dynamics in crystals: a macroscopic theory in a fractional Laplace setting*, Comm. Math. Phys. **333** (2015), no. 2, 1061–1105, DOI 10.1007/s00220-014-2118-6. MR3296170

[DPV17] Serena Dipierro, Stefania Patrizi, and Enrico Valdinoci, *Chaotic orbits for systems of nonlocal equations*, Comm. Math. Phys. **349** (2017), no. 2, 583–626, DOI 10.1007/s00220-016-2713-9. MR3594365

[DPV19] Serena Dipierro, Stefania Patrizi, and Enrico Valdinoci, *Heteroclinic connections for nonlocal equations*, Math. Models Methods Appl. Sci. **29** (2019), no. 14, 2585–2636, DOI 10.1142/S0218202519500556. MR4053239

[FIM09] Nicolas Forcadel, Cyril Imbert, and Régis Monneau, *Homogenization of some particle systems with two-body interactions and of the dislocation dynamics*, Discrete Contin. Dyn. Syst. **23** (2009), no. 3, 785–826, DOI 10.3934/dcds.2009.23.785. MR2461827

[GM12] María del Mar González and Regis Monneau, *Slow motion of particle systems as a limit of a reaction-diffusion equation with half-Laplacian in dimension one*, Discrete Contin. Dyn. Syst. **32** (2012), no. 4, 1255–1286, DOI 10.3934/dcds.2012.32.1255. MR2851899

[vGT14] L. I. van Griethuijsen and B. A. Trimmer, *Locomotion in caterpillars*, Biol. Rev. Camb. Philos. Soc. **89** (2014), no. 3, 656–670, DOI 10.1111/brv.12073.

[GB99] I. Groma and P. Balogh, *Investigation of dislocation pattern formation in a two-dimensional self- consistent field approximation*, Acta Mater. **47** (1999), no. 13, 3647–3654.

[HE73] S. W. Hawking and G. F. R. Ellis, *The large scale structure of space-time*, Cambridge Monographs on Mathematical Physics, No. 1, Cambridge University Press, London-New York, 1973. MR0424186

[HHW56] P. B. Hirsch, R. W. Horne, and M. J. Whelan, *Direct observations of the arrangement and motion of dislocations in aluminium*, Philosoph. Magazine **1** (1956), no. 7, 677–684, DOI 10.1080/14786435608244003.

[HCSW06] P. Hirsch, D. Cockayne, J. Spence, and M. Whelan, *50 years of TEM of dislocations: Past, present and future*, Philosoph. Magazine **86** (2006), no. 29-31, 4519–4528, DOI 10.1080/14786430600768634.

[Hea72] A. K. Head, *Dislocation group dynamics III. Similarity solutions of the continuum approximation*, Philosoph. Magazine **26** (1972), 65–72.

[IMR08] Cyril Imbert, Régis Monneau, and Elisabeth Rouy, *Homogenization of first order equations with (u/ε)-periodic Hamiltonians. II. Application to dislocations dynamics*, Comm. Partial Differential Equations **33** (2008), no. 1-3, 479–516, DOI 10.1080/03605300701318922.

[MP12a] Régis Monneau and Stefania Patrizi, *Derivation of Orowan's law from the Peierls-Nabarro model*, Comm. Partial Differential Equations **37** (2012), no. 10, 1887–1911, DOI 10.1080/03605302.2012.683504. MR2971210

[MP12b] Régis Monneau and Stefania Patrizi, *Homogenization of the Peierls-Nabarro model for dislocation dynamics*, J. Differential Equations **253** (2012), no. 7, 2064–2105, DOI 10.1016/j.jde.2012.06.019.

[Oro34a] E. Orowan, *Zur Kristallplastizität. I*, Zeit. für Physik **89** (1934), no. 9, 605–613, DOI 10.1007/BF01341478.

[Oro34b] E. Orowan, *Zur Kristallplastizität. II*, Zeit. für Physik **89** (1934), no. 9, 614–633, DOI 10.1007/BF01341479.

[Oro34c] E. Orowan, *Zur Kristallplastizität. III*, Zeit. für Physik **89** (1934), no. 9, 634–659, DOI 10.1007/BF01341480.

[PSV13] Giampiero Palatucci, Ovidiu Savin, and Enrico Valdinoci, *Local and global minimizers for a variational energy involving a fractional norm*, Ann. Mat. Pura Appl. (4) **192** (2013), no. 4, 673–718, DOI 10.1007/s10231-011-0243-9. MR3081641

[PS21] Stefania Patrizi and Tharathep Sangsawang, *From the Peierls-Nabarro model to the equation of motion of the dislocation continuum*, Nonlinear Anal. **202** (2021), Paper No. 112096, 50, DOI 10.1016/j.na.2020.112096. MR4138381

[PS] Stefania Patrizi and Tharathep Sangsawang, *preprint*.

[PV15a] Stefania Patrizi and Enrico Valdinoci, *Crystal dislocations with different orientations and collisions*, Arch. Ration. Mech. Anal. **217** (2015), no. 1, 231–261, DOI 10.1007/s00205-014-0832-z. MR3338445

[PV15b] Stefania Patrizi and Enrico Valdinoci, *Homogenization and Orowan's law for anisotropic fractional operators of any order*, Nonlinear Anal. **119** (2015), 3–36, DOI 10.1016/j.na.2014.07.010.

[PV16] Stefania Patrizi and Enrico Valdinoci, *Relaxation times for atom dislocations in crystals*, Calc. Var. Partial Differential Equations **55** (2016), no. 3, Art. 71, 44, DOI 10.1007/s00526-016-1000-0. MR3511786

[PV17] Stefania Patrizi and Enrico Valdinoci, *Long-time behavior for crystal dislocation dynamics*, Math. Models Methods Appl. Sci. **27** (2017), no. 12, 2185–2228, DOI 10.1142/S0218202517500427. MR3703556

[Pol34] M. Polanyi, *Über eine Art Gitterstörung, die einen Kristall plastisch machen könnte*, Zeit. für Physik **89** (1934), 660–664, DOI 10.1007/BF01341481.

[RCZ00] Paul H. Rabinowitz and Vittorio Coti Zelati, *Multichain-type solutions for Hamiltonian systems*, Proceedings of the Conference on Nonlinear Differential Equations (Coral Gables, FL, 1999), Electron. J. Differ. Equ. Conf., vol. 5, Southwest Texas State Univ., San Marcos, TX, 2000, pp. 223–235. MR1799055

[Sch07] Philip F. Schewe, *Polonium's simple cubic structure*, Physics Today **60(8)** (2007), no. 24, DOI 10.1063/1.4796556.

[RSJKEW03] R. Sedlacek, J. Kratochvi, and E. Werner, *The importance of being curved: bowing dislocations in a continuum description*, Philosophical Magazine **83** (2003), no. 31-34, 3735–3752.

[Sha07] Brahama D. Sharma, *Some elements go cubic under pressure*, Physics Today **60(10)** (2007), no. 17, DOI 10.1063/1.4797443.

[SH59] J. Silcox and P. B. Hirsch, *Direct observations of defects in quenched gold*, Philosoph. Magazine **4** (1959), no. 37, 72–89, DOI 10.1080/14786435908238228.

[Tay34a] G. I. Taylor, *The mechanism of plastic deformation of crystals. I: Theoretical*, Proc. R. Soc. Lond., Ser. A **145** (1934), 362–387, DOI 10.1098/rspa.1934.0106.

[Tay34b] G. I. Taylor, *The mechanism of plastic deformation of crystals. II: Comparison with observations*, Proc. R. Soc. Lond., Ser. A **145** (1934), 388–404, DOI 10.1098/rspa.1934.0107.

[Val06] Enrico Valdinoci, *Flatness of Bernoulli jets*, Math. Z. **254** (2006), no. 2, 257–298, DOI 10.1007/s00209-006-0947-5. MR2262703

[VSS06] Enrico Valdinoci, Berardino Sciunzi, and Vasile Ovidiu Savin, *Flat level set regularity of p-Laplace phase transitions*, Mem. Amer. Math. Soc. **182** (2006), no. 858, vi+144, DOI 10.1090/memo/0858. MR2228294

[Ver51] A. R. Verma, *Growth spiral patterns on carborundum crystals*, Nature **168** (1951), 783–784, DOI 10.1038/168783b0.

[Vol07] Vito Volterra, *Sur l'équilibre des corps élastiques multiplement connexes* (French), Ann. Sci. École Norm. Sup. (3) **24** (1907), 401–517. MR1509085

Department of Mathematics and Statistics, University of Western Australia, 35 Stirling Highway, Crawley WA 6009, Australia

Email address: serena.dipierro@uwa.edu.au

Department of Mathematics, The University of Texas at Austin, 2515 Speedway Austin Texas, 78712

Email address: spatrizi@math.utexas.edu

Department of Mathematics and Statistics, University of Western Australia, 35 Stirling Highway, Crawley WA 6009, Australia

Email address: enrico.valdinoci@uwa.edu.au